INTRODUCTION TO PROBABILITY AND STATISTICS

INTRODUCTION TO PROBABILITY AND STATISTICS

FROM A BAYESIAN VIEWPOINT

PART 1
PROBABILITY

BY

D. V. LINDLEY

Head of the Department of Statistics
University College London

CAMBRIDGE UNIVERSITY PRESS

CAMBRIDGE

LONDON NEW YORK NEW ROCHELLE

MELBOURNE SYDNEY

CAMBRIDGE UNIVERSITY PRESS
Cambridge, New York, Melbourne, Madrid, Cape Town, Singapore, São Paulo, Delhi

Cambridge University Press
The Edinburgh Building, Cambridge CB2 8RU, UK

Published in the United States of America by Cambridge University Press, New York

www.cambridge.org
Information on this title: www.cambridge.org/9780521055628

First published 1965
Reprinted 1969, 1976
First paperback edition 1980
Re-issued in this digitally printed version 2008

A catalogue record for this publication is available from the British Library

ISBN 978-0-521-05562-8 hardback
ISBN 978-0-521-29867-4 paperback

To

M. P. MESHENBERG

in gratitude

CONTENTS

PREFACE

The content of the two parts of this book is the minimum that, in my view, any mathematician ought to know about random phenomena—probability and statistics. The first part deals with probability, the deductive aspect of randomness. The second part is devoted to statistics, the inferential side of our subject.

The book is intended for students of mathematics at a university. The mathematical prerequisite is a sound knowledge of calculus, plus familiarity with the algebra of vectors and matrices. The temptation to assume a knowledge of measure theory and general integration has been resisted and, for example, the concept of a Borel field is not used. The treatment would have been better had these ideas been used, but against this, the number of students able to study random phenomena by means of the book would have been substantially reduced. In any case the intent is only to provide an introduction to the subject, and at that level the measure theory concepts do not appreciably assist the understanding. A statistical specialist should, of course, continue his study further; but only, in my view, at a postgraduate level with the prerequisite of an honours degree in pure mathematics, when he will necessarily know the appropriate measure theory.

A similar approach has been adopted in the level of the proofs offered. Where a rigorous proof is available at this level, I have tried to give it. Otherwise the proof has been omitted (for example, the convergence theorem for characteristic functions) or a proof that omits certain points of refinement has been given, with a clear indication of the presence of gaps (for example, the limiting properties of maximum likelihood). Probability and statistics are branches of applied mathematics—in the proper sense of that term, and not in the narrow meaning that is common, where it means only applications to physics. This being so, some slight indulgence in the nature of the rigour is perhaps permissible. The applied nature of the subject means that the student using this book needs to supplement it with some experience of practical data handling. No attempt has been made to provide such experience in the present book, because it would have made the book too large, and in any case other books that do provide it are readily available. The student should be trained in the use of various computers and be given exercises in the handling of data. In this way he will obtain the necessary understanding of the practical stimuli that have led to the mathematics, and the use of the mathematical results in understanding the numerical data. These two aspects of the subject, the mathematical and the practical, are complementary, and both are necessary for a full understanding of our subject. The fact that only one aspect is fully discussed here ought not to lead to neglect of the other.

The book is divided into eight chapters, and each chapter into six sections. Equations and theorems are numbered in the decimal notation: thus equation 3.5.1 refers to equation 1 of section 5 of chapter 3. Within §3.5 it would be referred to simply as equation (1). Each section begins with a formal list of definitions, with statements and proofs of theorems. This is followed by discussion of these, examples and other illustrative material. In the discussion an attempt has been made to go beyond the usual limits of a formal treatise and to place the ideas in their proper contexts; and to emphasize ideas that are of wide use, as distinct from those of only immediate value. At the end of each chapter there is a large set of exercises, some of which are easy, but many of which are difficult. Most of these have been taken from examinations papers, and I am grateful for permission from the Universities of London, Cambridge, Aberdeen, Wales, Manchester and Leicester to use the questions in this way. (In order to fit into the Bayesian framework some minor alterations of language have had to be made in these questions. But otherwise they have been left as originally set.)

The first part of the book, the first four chapters, is devoted to probability. The axioms of probability are stated in chapter 1 and elementary deductions made from them. In chapters 2 and 3 these results are applied to the study of random variables in one and higher dimensions respectively. Chapter 4 provides an introduction to the study of stochastic processes, including queueing theory, renewal theory and Markov chains.

The axiomatic structure used here is not the usual one associated with the name of Kolmogorov. Instead one based on the ideas of Renyi has been used. The essential difference between the two approaches is that Renyi's is stated in terms of conditional probabilities, whereas Kolmogorov's is in terms of absolute probabilities, and conditional probabilities are defined in terms of them. Our treatment always refers to the probability of A, given B, and not simply to the probability of A. In my experience students benefit from having to think of probability as a function of two arguments, A and B, right from the beginning. The conditioning event, B, is then not easily forgotten and misunderstandings are avoided. These ideas are particularly important in Bayesian inference where one's views are influenced by the changes in the conditioning event.

Another novelty in chapter 1 is the extensive discussion of probability as a degree of belief in section 6. The reason for this is that the treatment of statistics in the second part of the book is based on this form of probability. An attempt, at a modest mathematical level, has been made to justify the axiomatic structure for beliefs: in the hope that this will help to convince students of the reasonableness of the idea that beliefs can be measured numerically. The content of the first three chapters is a prerequisite for the second part on statistics. The study of stochastic processes in chapter 4 is not used in the latter part.

I am extremely grateful to J. W. Pratt, H. V. Roberts, M. Stone, D. J. Bartholomew; and particularly to D. R. Cox and A. M. Walker who made valuable comments on an early version of the manuscript; and to D. A. East who gave substantially of his time at various stages and generously helped with the proof-reading. Mrs M. V. Bloor and Miss C. A. Davies made life easier by their efficient and accurate typing. I am most grateful to the University Press for the excellence of their printing.

D.V.L.

Aberystwyth
April 1964

1

PROBABILITY

Introduction

The object of this book is to provide an introduction to the study of random phenomena. Most phenomena studied in science are not looked at from a random viewpoint but from a deterministic one, and may often be put into the simple form of cause and effect: if A then B. When randomness is present A may sometimes cause B, but sometimes C. Nevertheless, random phenomena, like deterministic ones, are capable of mathematical description; and in this book we show how this can be done using probability ideas. There are two aspects to the study. In the first half of the book, chapters 1–4, we deal with several simple random phenomena combining together to produce a more complicated random situation: the argument is purely deductive. In the second half, chapters 5–8, inductive problems are considered. If A_i always causes B_i, and no two of the B_i are the same, then it is a simple inductive process to infer that if B_i is observed then A_i must have been the cause. But if A_i can sometimes cause B_i and sometimes B_j, then the cause of B_j may be either A_i or A_j and the induction is not so complete. Nevertheless, again using probability ideas, inductive statements can be made precise. We shall often use the term 'statistics' to cover this latter study and reserve the term 'probability' for the purely deductive part.

Examples of deterministic statements of cause and effect are: if, under specified conditions, an electric current is passed through water, hydrogen and oxygen will be emitted; if the temperature of water is lowered sufficiently it will freeze; if a penny is released from the hand it will fall to the floor, and it can be calculated how fast it will fall. In contrast there are several situations in which it does not seem possible to make such statements of cause and effect, or if they were made they would be too complex to be useful. A simple example is provided by the release of a penny. If the penny is given a spinning motion it is still true that the penny will fall but it is not possible to say whether the effect will be that it will come to rest with the head or with the tail uppermost. Even if the initial conditions, or causes, were described with great care so that it would, in theory, be possible to predict the exposed face, it would not be useful to do so. Other, more practical, examples are commonplace. It is not possible to say whether or not an Englishman aged 40 will die within 10 years. It is not possible to say what the weather will be on a specified day a year hence. It is not possible to say what the hair colour of the child of brunette parents will be. It is not possible to say whether a given atom of uranium will disintegrate within 10 years. Here are several initial conditions but we cannot say what will be the corresponding effects. Although statements of the form, if A then B, cannot be made, important statements of a different kind can be made: witness, in the four examples quoted, the use of actuarial life tables, statements about climate, the science of genetics and the theory of radioactive decay.

Granted that it is possible to make precise statements about the toss of a penny (the way this can be done is described in the following sections) it is possible to deduce other results. For example, it can be calculated how long one can expect to go on tossing until the numbers of heads and tails are equal (§4.4); or how many heads one can expect to get in ten tosses (§2.1). This is a matter of mathematical deduction like deducing that the sum of the angles of a triangle is two right angles from the postulated properties of lines and points. From the random way in which telephone calls are made we can deduce the demand on an exchange to which a group of persons are connected (§§4.2, 4.3). We can answer questions such as what will be the effect of providing more telephone operators? These are deductive probability problems. Consider next a problem which is, in a sense, converse to the problem of the number of heads in ten tosses of a coin. If a penny has been observed to give eight heads in ten tosses is it likely to be a fair one obtained from a reputable mint? By fair we mean, to anticipate terms to be introduced later, that it is 'equally likely' to come down heads or tails in a single toss. From the fairness of a single toss the number of heads to be expected in ten tosses can be *deduced*, as just mentioned. To go in the reverse direction, from the results of the tosses to the fairness, means an *inductive* process, and is a problem typical of statistics. It is discussed in §7.2. Examples of more practical statistical problems are, to infer from agricultural experiments (which are subject to substantial random variation) the relative merits of different strains of wheat (§§6.4, 6.5); to compare the potencies of different drugs in an experiment on animals; to assess the accuracy of determination of a physical or chemical constant from an experiment which, like most experiments, must contain some randomness; and to detect and measure linkage between two genes.

1.1. The concept of a frequency limit

Consider a situation in which A can either cause an event B to occur or not, so that the situation is not deterministic. We say A either produces B or not-B, which we denote by $\bar{B}$. In many such situations it is possible to repeat A and observe, on each repetition, B or $\bar{B}$. It is an empirical fact that often, as the number, n, of repetitions increases, the ratio of the number, m, of times B occurs to the total number of repetitions becomes stable and appears to tend to a limit. Each repetition is termed a *trial*, the occurrence of the event B is a *success* (and of $\bar{B}$, a *failure*) and the ratio m/n the *success* (or *frequency*) *ratio*. The empirical observation can be expressed by saying that '$\lim_{n\to\infty}$' m/n exists, where the limit symbol has been placed in quotation marks to indicate that the notion is not the same as the ordinary mathematical limit. The limiting number is the empirical value of the *probability of B given A*, which is written $p(B|A)$. Any probability is a function of two arguments which are separated by a vertical line; the first is the event being considered, the second describes the conditions under which it is being considered and is called the *conditioning event*, or simply the *condition*: here, the event 'the occurrence of B' is being considered under condition A. In the next section axioms for probability suggested by this empirical fact will be given.

Empirical results

The simplest examples of repetitions of trials are provided by games of chance. If A denotes the toss of a penny and B denotes the fall of it with head uppermost, then repetitions are possible and the empirical fact observed. The limit is the probability of head on the toss of a coin. Kerrich (1946) carried out numerous experiments in order to demonstrate the stability of the success ratio: for example, he spun a coin 10,000 times and demonstrated that the ratio for 'heads' kept very near to 1/2.

If A denotes the roll of a die and B the occurrence of some number, or set of numbers, then again the empirical limit can be observed. Weldon rolled some dice and exhibited the stability of the success ratio for the event B, 'a 5 or 6'. The limit was not 1/3 as one might expect for a 'fair' die but somewhat more. It is relevant to notice that there is nothing in the above statement about the existence of the limit to say what the limit is. For a newly minted coin in our first example the limit is probably 1/2, but for a badly bent one it might well, as with Weldon's dice, be different from the ideal value.

The ultimate justification for probability does not lie in such experiments but rather in the practical success with which the theory has been applied. The prosperity of insurance houses is a witness to the value of statements about the probability of death. The science of genetics is based on the same sort of head or tail phenomenon as coin-tossing: will the gene transferred from parent to offspring be B or $b\,(=\bar{B})$? A breeding programme will provide the repetitions in which the stability can be observed, but this is secondary to the success of deductions from the theory. Games of chance provide the oldest example. The Chevalier de Méré played so many games that he was able, on empirical evidence alone, to detect that a certain limiting success ratio was less than 1/2: the mathematician Pascal showed that it was 0·491 by suitable application of probability theory. In radioactive studies the 'half-life' term is another way of expressing the limit: in repeated observations on different atoms, a success, namely decay, will have been observed in about 1/2 of them after the lapse of time equal to the 'half-life'.

The conditioning event

It is important to notice that the conditioning event A is just as relevant to the probability as is the event B. Consider tossing a coin where the event B, of success, is a head. Then, if the coin is a newly minted one, the probability will be about 1/2, but if the coin is badly bent, or if a piece of chewing-gum is stuck on to one face, the probability of the same event may be very far from 1/2. It is easy to produce paradoxes by failing to mention the conditioning event. Of course, in a series of trials it is impossible to keep the conditioning event completely constant. At least the time the trials are carried out will be different and typically the coin or die will show wear. But such difficulties of precision always arise in discussing the relationship between theory and practice: it is useful to think of an object as having a fixed weight, for example in use on a balance, but this is not precisely true. Conditions for the limit to exist within the theory will be given in §3.6 (theorems 2 and 3).

The form of 'lim'

The limit here is not a mathematical limit. That is to say, given any small positive number ϵ, it is not possible to find a value N such that $|m/n-p| < \epsilon$ for all $n > N$, where $p = \text{'lim'}_{n\to\infty} m/n$, as would be required of a mathematical limit. For there is nothing impossible in m/n differing from p by as much as 2ϵ, it is merely rather unlikely. And the word unlikely involves probability ideas so that the attempt at a definition of 'limit' using the mathematical limit becomes circular. The axiomatic approach (§1.2) avoids the difficulty, and the empirical observation will not be used to define probability, but only to suggest the axioms.

Examples

The probability statements that can be made in the examples used in the introduction are the following. The probability that an Englishman aged 40 will die within 10 years is 0·05: the event B is death within 10 years, A is the condition 'an Englishman aged 40'. The probability of rain at a specified place on a given date can be found from the rainfall statistics. The probability that the offspring of heterozygous parents will exhibit the recessive phenotype is $\frac{1}{4}$. The probability of decay of a uranium atom within $4{\cdot}49 \times 10^9$ years is $\frac{1}{2}$. All these results are based on the observation of repeated trials, supported, in the genetic case particularly, by additional indirect evidence.

1.2. The axioms of probability

The notion of an event is first formalized, and then the notion of the probability of an event. We consider a *sample space*, **A**, consisting of points, a, called *elementary events*. An *event* is a collection or *set* of elementary events and is denoted by a capital letter A, B, C, ..., with suffixes A_1, A_2, ..., where necessary. If a belongs to A we write $a \in A$. Selection of a particular a is referred to by saying 'a has occurred'. If $a \in A$ and a has occurred we say A has occurred. If A and B are two events, the set of a such that both $a \in A$ and $a \in B$ is denoted by AB. If AB has occurred then both A and B have occurred, and conversely. If $\{A_n\}$ is a sequence of events, the set of a which belong to at least one A_n is denoted by $\sum_n A_n$. If $\sum_n A_n$ has occurred then at least one A_n has occurred, and conversely. The members of the sequence are *exclusive given* C, if whenever C has occurred no two of them can occur together, that is if $A_m A_n C$ is the empty set whenever $m \neq n$. If the conditioning event C is **A**, the sample space, then, in this last definition, and similar ones, the words 'given **A**' are omitted. If $\sum_n A_n = \mathbf{A}$, that is, if every a belongs to at least one A_n, then $\{A_n\}$ is said to be *exhaustive*.

For certain pairs of events, A and B, a real number $p(A|B)$ is defined and called the *probability* of A given B. These numbers satisfy the following axioms:

Axiom 1. $0 \leqslant p(A|B) \leqslant 1$ and $p(A|A) = 1$.

Axiom 2. If the events in $\{A_n\}$ are exclusive given B then

$$p(\sum_n A_n | B) = \sum_n p(A_n | B).$$

Axiom 3 $p(C|AB)\,p(A|B) = p(AC|B)$.

The role of the axioms

The principle behind the construction of any axiom system is that a mathematical representation should be made of certain aspects of the real world. The elements in the mathematics are not parts of the real world but only representations of them. The elements are given properties (called axioms) supposed to reflect the behaviour of the corresponding parts of the real world. These properties can then be used in conjunction with the rules of mathematical logic to deduce other properties (within the mathematical system) which may be compared with the real world. If the axiomatization has been successful the comparison will lead to fruitful new ideas. The classical example is Euclid's geometry with axioms about lines and points, etc., such as 'through two points passes a unique line' and deductions like 'the sum of the angles of a triangle is two right angles'. Let us consider the axiom system described above in relation to the real phenomenon of tossing a penny.

The sample space

The event of a penny being tossed is more complicated than a mere occurrence of heads or tails: we could also consider its position of fall, the time of fall, etc., indeed countless other facets of the toss. Any particular toss may be represented by a point a and all possible tosses form the sample space. The collection of those a which result in heads is the event of heads having occurred. Any elementary event then will contain details of what penny was tossed, when and where it was tossed, etc. It is not necessary to be explicit in saying what is or is not contained in the description of an elementary event: it can be thought of quite abstractly as the toss being considered. It is often helpful to represent each elementary event by a point on the paper, and an event by a region of the paper containing the points representing the elementary events contained in the event. If A and B are two events so represented then the event $A+B$ is represented by the region which consists of the sum of the two regions for A and B. The event AB is represented by the region common to that for A and that for B. Thus in fig. 1.2.1 the event

A is represented by the horizontal rectangle: the event B by the vertical rectangle. $A+B$ is represented by the T-shaped figure and AB by the shaded square. Such a representation is called a *Venn diagram*. It will be found useful in understanding the proofs of the theorems in the next section.

Empirical justification for the axioms

Consider an event consisting of all tosses with a given coin under a standard set of conditions. Denote this event by B and the event of heads by A. Then $p(A|B)$ is the mathematical representation of the frequency limit discussed in §1.1. To see

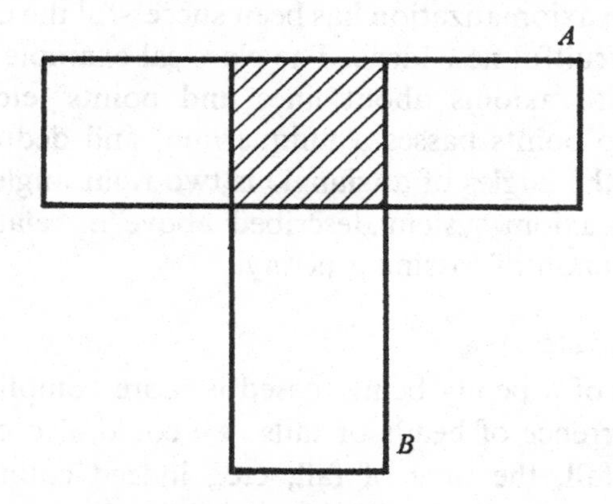

Fig. 1.2.1. Venn diagram for two events, A and B.

this, consider each axiom in turn. The frequency limit obviously lies between 0 and 1 and if we confine attention to tosses resulting in heads, so that the conditioning event is 'heads', then heads will always result and the limit necessarily be one, so that in any representation $p(A|A)$ must be one. This explains the first axiom. To appreciate the second consider N different events $A_1, A_2, \ldots, A_N$ concerning the outcome of the toss which are such that, given B, no two can occur together (they are exclusive, given B). For example, let A_i be the event that in the final rest position of the penny the acute angle between a fixed line on the head of the penny and a fixed line on the table lies between $(i-1)\pi/2M$ and $i\pi/2M$, the former limit being included and the latter excluded and M exceeding N. Then if, in n tosses,

A_i occurs m_i times, the event $\sum_i A_i$ (at least one A_i occurs) occurs $\sum_i m_i$ times. (This would not necessarily be true if the A_i were not exclusive.) The success ratio for $\sum_i A_i$ is $\sum_i m_i/n$, the sum of the success ratios for the individual events. Since the success ratios have this property it is reasonable to assume the same for their limits. Hence the second axiom. It is mathematically convenient to suppose this property holds for a countably infinite number of events† as well as for a finite number; the right-hand side will then be an infinite series, converging to the value on the left.

The first two axioms apply with a fixed conditioning event B. This is not so with the third axiom where two conditioning events B and AB both occur. To interpret this axiom let A and B be as before and let C be the event A_1, say, just referred to. Consider n tosses with the given penny under standard conditions, that is n occurrences of event B. Suppose there are m heads, A occurs m times. Suppose that amongst those m occasions C also occurs on r of them; so that r is the number of times the event AC occurs. Then trivially we have

$$\left(\frac{r}{m}\right)\left(\frac{m}{n}\right) = \left(\frac{r}{n}\right).$$

Now let $n \to \infty$, and hence $m \to \infty$ unless $p(A \mid B) = 0$. Then

$$\text{`lim'}_{m\to\infty} \frac{r}{m} = p(C \mid AB),$$

since the conditioning event is AB which occurs m times, and the observed event is C which occurs on r of these m occasions;

$$\text{`lim'}_{n\to\infty} \frac{m}{n} = p(A \mid B) \quad \text{and} \quad \text{`lim'}_{n\to\infty} \frac{r}{n} = p(AC \mid B) \quad \text{similarly.}$$

Consequently the third axiom is a reasonable limiting interpretation of the trivial result. If $p(A \mid B) = 0$, then necessarily $p(AC \mid B) = 0$ since AC is of rarer occurrence than A, and the result persists in that case. This completes the justification for

† The number of events (or other concepts) is countably infinite if they can be put into one-to-one correspondence with the integers 1, 2, ...

the axioms in the coin-tossing case: the reader might like to carry through one of the other examples similarly.

The reader may wonder why the complicated sample space of elementary events was introduced at all. Certainly, in simple problems like heads or tails in tossing pennies, it is not essential to have more than two distinct elementary events, 'heads' and 'tails' and it is often enough to use this sample space. But it is an advantage to consider the full sample space because then any event (such as C above) can also be discussed without changing the sample space. Had the sample space of only two elements been used, it would have had to have been altered before C could be discussed. Since the elementary events need not be formulated explicitly there is great advantage and little real addition in complexity in enlarging the sample space to its full extent. Furthermore, in defining the probabilities, it is only necessary to consider the events, and not the elementary events, so that the introduction of complicated elementary events does not increase the difficulty of defining the probabilities. Thus, in coin tossing, however involved the description of the individual tosses be, only the events of heads and tails need be considered.

Fixed conditions

It often happens throughout a probability calculation that one event B always occurs as part or the whole of the conditioning event: for example, the event B just defined as tosses with a single coin under standard conditions. B is often the whole sample space, $\mathbf{A}$, or may be taken to be that (see the comment on theorem (1.4.1) below). It then economizes on notation to write $p(A)$ for $p(A|\mathbf{A})$ and $p(A \mid C)$ for $p(A \mid \mathbf{A}C)$, omitting reference to $B = \mathbf{A}$. Indeed most writers on probability define probability as $p(A)$, a function of a single event: but this can be misleading and it pays to remember the conditioning event. To omit it is rather like considering an effect without its cause. (Compare the discussion in §1.1.)

One method of constructing a probability system is as follows. With certain events, A, associate a real number $p(A)$ satisfying

Axiom 1*a*. $0 \leqslant p(A) \leqslant 1$ and $p(\mathbf{A}) = 1$.

Axiom 2*a*. If the events in $\{A_n\}$ are exclusive, given A, then

$$p(\sum_n A_n) = \sum_n p(A_n).$$

Define $p(A \mid B)$ by $p(A \mid B) = p(AB)/p(B)$ unless $p(B) = 0$ when it is undefined. Then it is easy to verify that $p(A|B)$ satisfies axioms 1–3. Most probability systems can be constructed this way but there are exceptions (see example 3, below). Notice that $p(A) = p(A \mid \mathrm{A})$. $p(A)$ is sometimes called an *absolute* probability and $p(A \mid B)$ a *conditional* probability: but all probabilities are really conditional ones.

Notice that we have not stated for which events $p(A|B)$ is defined: we have merely said 'for certain pairs of events, A and B, ...'. At the level of mathematical discussion used in this book we shall scarcely ever have any trouble over this matter. The only difficulty we shall encounter is that sometimes $p(A|B)$ cannot be defined because $p(B) = 0$. This is especially so in the construction of a probability system using axioms 1*a* and 2*a* wherein $p(A \mid B) = p(AB)/p(B)$. Within the full axiom system this point does not arise so often: an illustration is given in example 3 below where $p(B)$, in the notation of that example, is zero yet $p(A_j|B)$ is defined. Readers with the requisite mathematical knowledge may recognize that the observed events A will have to belong to a *Borel field* or *σ-algebra*. The conditioning events will have to belong to a subcollection of the Borel field. A complete discussion is given by Renyi (1962), or using axioms 1*a* and 2*a* by Loève (1960) and many others.

Odds

It is sometimes convenient to use another language to describe probabilities. This is the language favoured by bookmakers and will be used in §1.6 when we discuss a different type of probability from that based, as here, on frequency considerations. If A_1 and A_2 are two exclusive events (usually $A_2 = \bar{A}_1$) then the *odds* on A_1 against A_2, given B, is the ratio $p(A_1 \mid B)/p(A_2 \mid B)$ to 1. Odds do not usually have such simple properties as probabilities but sometimes, especially in using Bayes's theorem (1.4.6), the ratio occurs more naturally than the separate probabilities. An illustration occurs in example 1.4.2. If $A_2 = \bar{A}_1$ and $p(A_1|B) = p$, then the odds are $p/(1-p) = b$, say, to 1 and $p = b/(1+b)$, so that the odds and probabilities are equivalent in the sense that one can be found from the other.

Examples

It is remarkable that these three quite simple axioms are enough to build up the whole structure of probability theory with the aid only of mathematical logic; but it is so, and the rest of this book will demonstrate it. There is an advantage in keeping the axioms few and simple because it becomes an easy matter to check whether any system is a probability system by seeing if it satisfies the axioms. If it does, any probability theorem may immediately be applied to this system. We now give four examples of probability systems.

Example 1. Let **A** be the unit interval of real numbers a. The method of axioms 1a and 2a is used to construct a probability system. Let A be any interval contained therein: that is A is the set of numbers a satisfying $a_1 \leqslant a \leqslant a_2$ where $0 \leqslant a_1 \leqslant a_2 \leqslant 1$. Define $p(A)$ as the length $(a_2 - a_1)$ of the interval A. Let other sets have probabilities defined by axiom 2a. It is clear that $p(A)$ satisfies those two axioms and consequently $p(A \mid B) = p(AB)/p(B)$ satisfies axioms 1–3, provided B has not zero length, when $p(A|B)$ is undefined.

Two important points emerge from this example. First, length has basically the same mathematical properties as absolute probability—indeed it is a special case. Probability is therefore often called, for a fixed conditioning event, a *measure*; length is a particular measure. It will often be convenient in subsequent developments to use an interpretation of probability in terms of length, area or volume. Secondly, we see how the brevity and simplicity of the axioms make it easy to verify whether or not any mathematical set-up can be interpreted as a probability.

The example might be relevant if an experimenter was measuring an angle. If θ is the angle in radians with a fixed direction, then $a = \theta/2\pi$ lying between a_1 and a_2 would correspond to the occurrence of the event A above. If all angles were equally likely (in the colloquial use of the term) then $p(A)$ is as in the example, and the mathematical set-up is an interpretation of the colloquialism. Roulette is a possible situation where all angles are equally likely: if A is the event 'black' and B the event 'even' then $p(A \mid B)$ is the probability of 'black' given that the ball is in a compartment with an even number.

Example 2. Suppose that, in a single trial, there are only a finite number of possible exclusive events, $A_1, A_2, \ldots, A_k$. These may be taken as the elementary events and the sample space contains k points. Let $p_1, p_2, \ldots, p_k$ be non-negative numbers with $\sum_i p_i = 1$. It is easy to verify that with $p(A_i) = p_i$, axioms $1a$ and $2a$ are satisfied and hence a probability system can be constructed. It is called the *finite discrete* case. In the case $k = 2$, $A_2 = \bar{A}_1$ and the event A_1 is observed either to occur or not. $k = 6$ is relevant when rolling a die and for a 'true' die $p_i = 1/6$ for all i. The sample space consists of k points and p_i is often called the probability *mass* at the point. Indeed an analogy with mechanics is possible in both examples 1 and 2. The first may be compared with a uniform rod of unit length: the second with a number of particles of masses proportional to the probability masses and situated at the points in sample space. The notions of centre of gravity and moment of inertia have probability analogues (see §§2.1, 2.2, 2.4).

Example 3. The former example extends to a countably infinite number of exclusive events, A_i, with $\{p_i\}$ a sequence of non-negative numbers whose sum converges to one. Whilst this covers a wide class it does not include the case where all events have the same probability, because an infinity of numbers p_i cannot be all equal and have sum 1. To include this important case the full axiom structure has to be used; the probability system cannot be defined by absolute probabilities as in the other examples. Instead we define $p(A_j|B)$ whenever $B = \sum_{i=1}^{n} A_m$ (the summation being over any *finite* number, n, of events) as n^{-1} if A_j is one of A_{m_i} and 0 otherwise. We leave the reader to check that this defines a probability system satisfying axioms 1–3. The definition of $p(A_j|B)$ means that amongst any finite collection, B, all events are equally likely.

Example 4. Suppose an actuary is considering the insurance of lives of Englishmen. Then he will formally be considering a sample space of all Englishmen alive at a given date, and each elementary event will correspond to one Englishman. Part of the description of a man is his age and consequently he will be able to talk of the event of being over 40 as that event which contains only those elementary events having that property. If only this event is of interest then the finite discrete case of example 2 is relevant with $k = 2$, A_1 being over 40 and $A_2 = \bar{A}_1$, and probability may be defined as in that example. But he may also wish to consider the men's heights, weights, blood-sugar contents, etc., and more complicated events have to be considered. We shall have to leave the problem of defining probabilities for such events until chapters 2 and 3.

1.3. Independence

Two events, A and B, are said to be *independent given C* if

$$p(AB \mid C) = p(A \mid C)\, p(B \mid C). \tag{1}$$

This is a particular case of the following important

Definition. *The events, finite or infinite, of a sequence $\{A_n\}$ are* independent *given B if, for* any *finite collection of them, say $A_{n_1}, A_{n_2}, \ldots, A_{n_k}$, the equation*

$$p(A_{n_1} A_{n_2} \ldots A_{n_k} \mid B) = p(A_{n_1} \mid B)\, p(A_{n_2} \mid B) \ldots p(A_{n_k} \mid B) \tag{2}$$

obtains.

$A_{n_1} A_{n_2} \ldots A_{n_k}$ denotes the event which occurs iff† all the events $A_{n_1}, A_{n_2}, \ldots, A_{n_k}$ occur: it is an extension of the notation AB used in §1.2. If $B = \mathbf{A}$ we sometimes omit reference to the conditioning event (cf. §1.2) and speak of the events being independent, writing $p(A_{n_1} \mid \mathbf{A}) = p(A_{n_1})$, etc.

Justification for the definition

The motivation behind the first definition is that in some circumstances the occurrence of B may not affect the probability of occurrence of A. Suppose C has occurred and is the conditioning event; then $p(A \mid C)$ measures the frequency of occurrence

† 'iff' means 'if and only if'.

of A in these conditions. Suppose one is then told that, in addition to C, another event B has occurred: then the conditioning event is BC and one appears to have been given more information than originally when one was told only that C had occurred. Or, to look at it another way, BC contains certainly no more, and typically fewer, elementary events than C. The new frequency of A is measured by $p(A \mid BC)$. Now it may happen that B in fact provides no new information about the probability of A, that is

$$p(A \mid BC) = p(A \mid C), \tag{3}$$

in which case it is reasonable to describe A as independent of B given C. As it stands this definition is not symmetric in A and B but if each side of (3) is multiplied by $p(B \mid C)$, and axiom 3, which in the notation of this section reads

$$p(A \mid BC)\, p(B \mid C) = p(AB \mid C),$$

is used on the left-hand side, we obtain (1) and use this as a (symmetric) definition of independence.

An example of independent events is provided by considering the sample space consisting of simultaneous tosses of two coins: under reasonable conditions the events 'head for first coin' and 'head for second coin' will be independent (given **A**). An example of non-independent events is provided by considering the sample space of all possible weather conditions at a fixed place over two consecutive days: the events 'rain on the first day', A, and 'rain on the second day', B, are not independent (given **A**), for it is an empirical fact that $p(B \mid A) > p(B)$, since consecutive days tend to be alike. Independence of transmission of two genes, and non-linkage of them, are two expressions of the same idea. For if one gene is linked with another the occurrence of one in an offspring increases the probability of occurrence of the other: this does not happen if they are not linked.

The extension to more than two events requires care. If a set of events is to be independent it should mean that if *any* number of them are known to have occurred the probability of any other one of them should be unaltered: in symbols

$$p(A_{n_1} \mid A_{n_2} \dots A_{n_k} B) = p(A_{n_1} \mid B). \tag{4}$$

The symmetrical form of this is equation (2). It is enough in most applications to use only finite collections. Notice that, even for a finite set of N events, it is not enough that

$$p(A_1 A_2 \dots A_N \mid B) = p(A_1 \mid B)\, p(A_2 \mid B) \dots p(A_N \mid B), \qquad (5)$$

for this does not imply, for example, when $N > 2$,

$$p(A_1 A_2 \mid B) = p(A_1 \mid B)\, p(A_2 \mid B), \qquad (6)$$

so that, when this equation is violated, A_2 on its own may give information about A_1. Equations (2) and (4) must hold for all finite collections. The phrase *stochastically independent* is sometimes used to distinguish the concept from that of functional independence. It is usually clear which is intended and only in cases of doubt will the qualification be inserted. Stochastic is a synonym for probability (as an adjective).

Random trials

An important case of independence arises in certain types of trial (§1.1). Each trial can result in success or failure. It is convenient to represent a success by 1 and a failure by 0. Then the outcome of a sequence of trials can be represented by a sequence (finite or infinite according to the number of trials) of 0's and 1's, and each elementary event is such a sequence. Let A_i be the event, a 1 in the ith place, or equivalently a success in the ith trial. If the A_i are independent (given A) we say that the sequence of trials is *random*: the result of one trial does not affect probabilities in any other trial. Coin tossing provides an example. A random sequence of trials with $p(A_i) = p$ for all i is called a *random sequence with constant probability of success*,† and it is to such a sequence that the empirical limit of the success ratio applies. In §3.6 a result will be proved which represents, within the mathematics, the empirical fact of the existence of the limit. Notice that the limit may exist when the trials are not independent; for example, where the trials are days and failure is rain; then climate is an expression of the limit but, as already mentioned, successive days are not independent.

† Sometimes it is called a *Bernoulli* sequence.

Populations

A second example of independence arises with groups of people, animals, insects, etc. Such a group is termed a *population.* Let a population contain N members, and consider a sample space of N^n points, each point consisting of n members of the population arranged in order, repetitions being allowed, no two points being the same. Define a probability system by letting each elementary event have probability N^{-n}. This is a finite discrete sample space (example 2 of § 1.2). Let A_{ij} be the event that the jth member in the order be the ith member of the population. Then clearly $p(A_{ij}) = N^{-1}$ and A_{ij} and A_{ik} are independent for $j \neq k$. Each elementary event is termed a *random sample of size n with replacement* from the population. In practice this example arises when one member of the population is taken, inspected and returned to the population; then a second member taken, and so on; where, at any stage, each member has the same chance (or probability) of being taken, namely N^{-1}, and n members are taken. Such a process is called random sampling *with replacement.* It is more usual to sample *without replacement*, i.e. the taken member is not returned to the population. Here the sample space contains only the $N(N-1) \dots (N-n+1) = N!/(N-n)!$ arrangements of *n different* members in order, and, if each member left has the same chance of being taken, each elementary event has probability $(N-n)!/N!$. Clearly $p(A_{ij}) = N^{-1}$ but $p(A_{i2} | A_{j1}) = (N-1)^{-1}$ for $i \neq j$ because the second member cannot be j, so there are only $(N-1)$ left to choose from. Hence A_{i2} and A_{j1} are not independent. Each elementary event is a *random sample of size n without replacement*: this being the more common form it is often called simply a *random sample.* Such a sampling technique is sometimes used with human populations in social studies but it is often difficult to ensure that all members have the same chance of appearing in the sample, and more complicated methods are used.

If N is large compared with n, that is a small sample is taken, then the departure from independence due to the non-replacement is not every great as $(N-1)^{-1}$ is little different from N^{-1}. Even when the last member of the sample is about to be taken each member left in the population has only probability $(N-n+1)^{-1}$ of being taken, again little different from N^{-1}. Often N is taken to be infinite. The mathematical model for this is more complicated and cannot be constructed from absolute probabilities as described in §1.2 and used here for finite N. The method of example 3, §1.2, has to be used. The elementary events are arrangements of n different members of the population. Let B be any set of k (finite) elementary events. Then $p((a)\,|\,B) = k^{-1}$ if $a \in B$ and zero otherwise, where (a) is the event consisting of the single elementary event a. This defines a random sample of size n from an infinite population. The reader can verify that $p(A_{i2}\,|\,A_{j1}B) = p(A_{i2}\,|\,B)$ whenever B is finite and symmetric in a finite number of members including i and j, $i \neq j$; that is, conditional on a finite number of members of the population being used to form the sample. Hence given B, A_{i2} and A_{j1} are independent.

If a proportion p of the members of a population possess some character, then the random sampling may be thought of as n independent trials with probability p of success (an individual possessing the character). If the population is finite this ignores the dependence due to nonreplacement of members in the population; but is exact for infinite populations and is a reasonable approximation for large populations. Hence a probability may often be thought of as a proportion in a population. This is useful in genetics. The relationship between 'probability of rain tomorrow' and 'proportion of rainy days' is similar, though here the population of days is not as real as an animal population. Some authors use hypothetical populations (usually infinite) but this is unnecessary. Populations need never be used but the idea is often useful. When they really exist (as with animals) then they can be represented, as just described, within the axiom system.

The idea of the last paragraph may be generalized from the case where each member does, or does not, possess a characteristic, to the case where each member has associated with him a real number; for example, his height (see the discussion on histograms in §2.4). Random samples of this type are basic in statistical arguments (see §5.1).

1.4. Bayes's theorem

The second axiom is often called the *addition law* of probabilities. Similarly, the independence condition (1.3.2) is often called the *multiplication law* of probabilities. Notice that the addition law only holds for exclusive events and the multiplication law for independent events. We now deduce, from the axioms, other simple but useful results. Where they do not lose their content by doing so, the theorems will be stated and proved with $\mathbf{A}$ as the conditioning event. $p(A|\mathbf{A})$ will be written $p(A)$. The general result for any conditioning event B can be obtained by inserting 'given B' at obvious places (for example, after 'independent'), inserting B after the vertical line and replacing $p(.)$ by $p(.|B)$. The advantage in only using $\mathbf{A}$ is that the results are more easily understood. Notice that with $B = \mathbf{A}$ the third axiom reads

$$p(C|A)\,p(A) = p(CA). \tag{1}$$

Theorem 1. $p(A|B) = p(AB|B)$.

This follows from the third axiom which says that

$$p(AB|B) = p(A|BB)\,p(B|B)$$

which equals $p(A|B)$ since $BB = B$ and $p(B|B) = 1$ by the first axiom.

An event A is said to *imply*, given C, an event B if, given C, whenever A occurs, B occurs: alternatively, every $a \in AC$ also belongs to B; or the part of the set A which is in C is part of the set B. Then $ABC = AC$, or if $C = \mathbf{A}$, $AB = A$. A Venn diagram may help the reader (§1.2).

Theorem 2. *If A implies B*

$$p(A|B)\,p(B) = p(A).$$

This follows, with a change of notation (A for C, B for A), from (1) with the condition $AB = A$.

Theorem 3. *If A implies B, $p(A) \leqslant p(B)$.*

We have, from theorem 2, $p(A) = p(A|B)\,p(B)$; but $p(A|B) \leqslant 1$ by the first axiom and the result follows.

Theorem 4 *(Generalized addition law). If $\{A_n\}$ are exclusive and exhaustive and B is any event, then*

$$p(B) = \sum_n p(B \mid A_n)\, p(A_n).$$

By (1) each term on the right-hand side is $p(BA_n)$. Since the $\{A_n\}$ are exclusive and exhaustive the $\{BA_n\}$ are exclusive and $\sum_n A_n B = B$. The result follows from the addition law.

Theorem 5. *If $p(B)$ does not vanish then*

$$p(A \mid B) = p(B \mid A)\, p(A)/p(B).$$

This is immediate from (1).

Theorem 6 *(Bayes's theorem). If $\{A_n\}$ is a sequence of events and B is any other event with $p(B) \neq 0$, then*

$$p(A_n \mid B) \propto p(B \mid A_n)\, p(A_n).$$

This follows from theorem 5 on putting A_n for A. The missing constant of proportionality is $p(B)^{-1}$.

Corollary. *If, in addition, the $\{A_n\}$ are exclusive and exhaustive then*

$$p(A_n \mid B) = p(B \mid A_n)\, p(A_n)/\sum_i p(B \mid A_i)\, p(A_i).$$

This is immediate on rewriting $p(B)$ as a summation, by the generalized addition law (theorem 4).

Theorem 7. *If $\{A_n\}$ are exclusive and not all $p(A_n) = 0$, and B is any event, then*

$$p(B \mid \sum_n A_n) = \sum_n p(B \mid A_n)\, p(A_n)/\sum_n p(A_n).$$

The left-hand side is

$$p(B \sum_n A_n)/p(\sum_n A_n) = \sum_n p(BA_n)/\sum_n p(A_n),$$

on application of the addition law to numerator and denominator; which is valid since $\{A_n\}$ and hence $\{BA_n\}$ are exclusive. The result follows since $p(BA_n) = p(B \mid A_n)\, p(A_n)$.

***Theorem 8** (Generalized multiplication law). If $A_1, A_2, \ldots, A_n$ are any events and the event $A_1 A_2 \ldots A_{n-1}$ has not zero probability, then*

$$p(A_1 A_2 \ldots A_n) = p(A_1)\, p(A_2 \mid A_1)\, p(A_3 \mid A_1 A_2) \ldots p(A_n \mid A_1 A_2 \ldots A_{n-1}).$$

The result is immediate by repeated use of (1).

The reader should follow through the simple proofs if only to convince himself that our results do follow from the simple axioms. Theorem 1 is a reflexion of the fact that, given B, only elementary events in B need be considered and A may be replaced by AB: elementary events in A but not in B are irrelevant. This explains why, if one conditioning event occurs throughout a calculation, reference to it can be omitted. For, if B always occurs, since $p(AB \mid B) = p(A \mid B)$, only that part of A in B need be considered in place of A. It is unnecessary to take account of A without B. Theorem 2 is merely a special case of (1) and is used, for example, in deriving equation 1.5.7. Theorem 3 is useful when proving that a sequence of probabilities $\{p(A_n)\}$ tends to zero. If A_n implies B_n it may be easier to prove $p(B_n)$ tends to zero. Wide use is made of theorem 4 in calculating $p(B)$ when B can happen in several ways, corresponding to the A_n. The following example illustrates a use of theorem 4.

Example 1. The sample space is the space of all human twins (cf. §1.3). A pair of twins can either be identical, an event denoted by A, or they can be fraternal, which is the complementary event to A and therefore denoted by $\bar{A}$. Similarly, the pair can either be both male, MM, both female, FF, or of mixed sex, MF. The problem is to estimate empirically the probability of a pair of twins being identical: in other words the proportion of identical twins amongst the population of twins. This is somewhat difficult to do directly because it would involve determining whether or not a given pair was identical; a difficult, if not impossible, procedure. However, the probability of both members being male can easily be found, and using the connexions between the events MM, MF and FF on the one hand, and the event A on the other, the unknown $p(A)$ can be found. We shall assume these connexions to be of the following form:

$$p(MM \mid A) = p(FF \mid A) = \tfrac{1}{2}, \quad p(MF \mid A) = 0;$$

$$p(MM \mid \bar{A}) = p(FF \mid \bar{A}) = \tfrac{1}{4}, \quad p(MF \mid \bar{A}) = \tfrac{1}{2}.$$

These equations express the facts that males and females are equally likely and that the two members of a pair of fraternal twins have independent determinations of sex (since $p(MM \mid \bar{A}) = p(M \mid \bar{A})\, p(M \mid \bar{A})$, etc.). Actually it is known that the sex ratio of male to female at birth is a little over one but this will only affect the arithmetical details. Now theorem 4 says that in this situation

$$p(MM) = p(MM \mid A)\, p(A) + p(MM \mid \bar{A})\, p(\bar{A}),$$

and, inserting the known values,

$$p(MM) = \tfrac{1}{2}p(A) + \tfrac{1}{4}(1 - p(A)).$$

Whence $$p(A) = 4p(MM) - 1.$$

$p(MM)$ can be found from observation and hence $p(A)$.

Theorem 5 is useful in calculating $p(A \mid B)$ from $p(B \mid A)$ and in the form given in theorem 6 is the most important result in the book, at least for the inductive, statistical applications. Rev. Thomas Bayes (1702–61) had his famous paper published after his death, in 1763. The paper has recently been republished (1958) because of its great importance. Virtually the whole of the second part (chapters 5–8) of this book is an application of his result and here we only give two simple uses of it.

Example 2. A machine producing articles can go wrong in a number of ways: when it does it may produce defective articles. The problem is to determine what is wrong with the machine from inspection of the nature of the defect. Each elementary event concerns a particular breakdown, with which may be associated a particular defective article. A_i $(1 \leqslant i \leqslant m)$ is the event that the breakdown is of the ith type. B_j $(1 \leqslant j \leqslant n)$ is the event that the article's defect is of the jth type. It is assumed that neither can the machine break down nor the article be defective in more than one way at the same time ($\{A_i\}$ and $\{B_j\}$ are both collections of exclusive events) and that a defective article is necessarily a product of a broken machine. The breakdown rates (the $p(A_i)$) are known from past experience, as are the proportions of the different types of defective articles produced by each type of breakdown (the $p(B_j \mid A_i)$). Theorem 6 enables $p(A_i \mid B_j)$ to be calculated. Then if a defect of type j is observed (that is B_j is the conditioning event) the probability that the machine breakdown is of type i is known: in particular if one A_i has probability near one, then that type of breakdown would be the one suspected. Consider a numerical example with $m = n = 2$: suppose

$$p(A_1)\dagger = \tfrac{1}{3}, \qquad p(A_2) = \tfrac{2}{3};$$
$$p(B_1 \mid A_1) = \tfrac{1}{2}, \quad p(B_2 \mid A_1) = \tfrac{1}{10};$$
$$p(B_1 \mid A_2) = \tfrac{1}{10}, \quad p(B_2 \mid A_2) = \tfrac{1}{4}.$$

Notice that B_1 and B_2 are not exhaustive: a breakdown need not cause a defect. We use the theorem first for B_1.

If B_1 occurs:

$$p(A_1 \mid B_1) \propto \tfrac{1}{2} \times \tfrac{1}{3} = \tfrac{1}{6},$$
$$p(A_2 \mid B_1) \propto \tfrac{1}{10} \times \tfrac{2}{3} = \tfrac{1}{15},$$

so that the odds are 5 to 2 ($\tfrac{1}{6}$ divided by $\tfrac{1}{15}$) on A_1 against A_2, although, without knowledge of B_1, A_1 only occurs half as often as A_2. The mechanic should suspect that A_1 is the cause of the observed defect.

If B_2 occurs:

$$p(A_1 \mid B_2) \propto \tfrac{1}{10} \times \tfrac{1}{3} = \tfrac{1}{30},$$
$$p(A_2 \mid B_2) \propto \tfrac{1}{4} \times \tfrac{2}{3} = \tfrac{1}{6},$$

so that the odds are 5 to 1 on A_2 against A_1: hence A_2 is suspected as the cause. Notice that it is not necessary to calculate the constant of proportionality; it would be easy to do so. The probabilities can be determined from the condition that, since the $\{A_i\}$ are exhaustive given a defect, their sum must be one.

The above is a fairly typical example of the use of Bayes's theorem. The main points to notice are the presence of a number of possibilities (the $\{A_i\}$) of which one, and only one, must obtain, but it is not known which; and some event which has

† That is $p(A_1 \mid A)$.

been observed to occur (one of the $\{B_j\}$) and which is influenced by the $\{A_i\}$ (through $p(B_j|A_i)$). We wish to state what we know about which unknown possibility occurred (the $p(A_i|B_j)$). Examples are, where $\{A_i\}$ correspond to different diseases and B_j is an observed symptom; where $\{A_i\}$ are the possible chemical structures an unknown product might have and B_j is the result of a chemical analysis; where $\{A_i\}$ correspond to different scientific theories that are available to explain a phenomenon and B_j is an experimental result; where $\{A_i\}$ are the suspects and $\{B_j\}$ the clues in a detective novel.

The more important applications of Bayes's theorem require an extension of the concept of probability, which is given in §1.6. For the moment we content ourselves with a second example in which the main interest lies in the language used in the original formulation and the 'translation' of it into probability language.

Example 3. Particles of s types are in proportions $p_1, p_2, \ldots, p_s$. They meet a barrier and the probability of the rth type of particle passing through the barrier is q_r. Find the proportions of the s types after passage through the barrier.

The sample space is obviously made up of elementary events consisting of particles arriving at the barrier. Let A_r be the set of all elementary events in which the particle is of type r: then $p(A_r) = p_r$ (cf. §1.3 again). Let B be the set of all elementary events in which the particle passes through the barrier. Then $p(B|A_r) = q_r$. Now $p(A_r|B)$ means the probability that the particle is of type r given that it has passed the barrier, which is the required proportion. So Bayes's theorem says

$$p(A_r|B) = \frac{q_r p_r}{\sum_i q_i p_i},$$

the required result.

An example of the use of theorem 7 will be given at the end of §1.5. The principal use of theorem 8 is to derive the probability of several events occurring (left-hand side) in terms of the conditional probabilities of each event separately (right-hand side). The following example illustrates this and also the formulation of a probability system of trials which are not independent.

Example 4. A sequence of trials is considered, each of which, except the first, can result in a 0 or a 1. The first trial always results in a 0. In any subsequent trial the probability of a result 1, given that the previous trial resulted in a 0, is p independently of the results of all earlier trials; and the same statement is true with 0 and 1 interchanged. In other words, the probability of a change of the values in consecutive trials is p. We write $q = 1-p$. This is a model for a random telegraph signal, the signal being provided by the change in the result. Let A_i be the event, a 1 on the ith trial, and, as before, $\bar{A}_i$ be the event, a 0 on the ith trial. Then the probability of the sequence 0011 is, by theorem 8,

$$\begin{aligned} p(\bar{A}_1\bar{A}_2A_3A_4) &= p(\bar{A}_1)\,p(\bar{A}_2\,|\,\bar{A}_1)\,p(A_3\,|\,\bar{A}_1\bar{A}_2)\,p(A_4\,|\,\bar{A}_1\bar{A}_2A_3) \\ &= p(\bar{A}_1)\,p(\bar{A}_2\,|\,\bar{A}_1)\,p(A_3\,|\,\bar{A}_2)\,p(A_4\,|\,A_3), \end{aligned}$$

because of the independence of the result of a trial from the results of all trials other than the immediately preceding one. Hence

$$p(\bar{A}_1\bar{A}_2A_3A_4) = 1.q.p.q = pq^2.$$

Notice that the trials are not independent, unless $p = \frac{1}{2}$, because otherwise, $p = p(\bar{A}_3\,|\,A_2) \neq p(\bar{A}_3\,|\,\bar{A}_2) = q$. A sequence of trials in which, as here,

$$p(A_n\,|\,A_{n-1}C_{n-2}) = p(A_n\,|\,A_{n-1}),$$

where C_{n-2} is any result of the first $(n-2)$ trials, is called a *Markov* sequence after the Russian mathematician of that name. They will be discussed in §§4.5 and 4.6.

1.5. Genetical applications

An important field of application of probability theory is genetics, the basic Mendelian laws being essentially probabilistic. Throughout this section some familiarity with elementary genetical ideas will be assumed. The results will not be used outside this section.

Example 1. First consider the case of a single gene, with alleles A and a, for which the three genotypes AA, Aa, aa give rise to three distinct phenotypes, and suppose that the proportions of the three phenotypes in the population are initially p_0, q_0, r_0 respectively. If mating takes place randomly, we are concerned with random samples of size 2 (cf. §1.3) from the population, supposed infinite, and the six possible types of mating with their probabilities are:

$$\left.\begin{matrix} AA\times AA & AA\times Aa & AA\times aa & Aa\times Aa & Aa\times aa & aa\times aa \\ p_0^2 & 2p_0q_0 & 2p_0r_0 & q_0^2 & 2q_0r_0 & r_0^2 \end{matrix}\right\}. \quad (1)$$

(The factor 2 in the probabilities of mixed matings arises because $AA\times Aa$, for example, is identical with $Aa\times AA$, each having probability p_0q_0.) For any type of mating (corresponding to the conditioning event) the conditional probabilities of the offspring follow from Mendel's laws. Thus, if Aa also denotes the event of an offspring being of type Aa, $p(Aa\,|\,AA\times Aa) = \frac{1}{2}$, etc. The proportions of phenotypes in the next generation, $p(AA)$, etc., can be found from the generalized addition law. For example,

$$\begin{aligned} p(AA) = {} & p(AA\,|\,AA\times AA)\,p(AA\times AA) \\ & + p(AA\,|\,AA\times Aa)\,p(AA\times Aa) \\ & + p(AA\,|\,Aa\times Aa)\,p(Aa\times Aa), \end{aligned} \quad (2)$$

the remaining conditional probabilities being zero. Hence

$$p(AA) = 1\times p_0^2 + \tfrac{1}{2}\times 2p_0q_0 + \tfrac{1}{4}\times q_0^2 = (p_0 + \tfrac{1}{2}q_0)^2. \quad (3)$$

This result can be obtained more easily by noticing that the proportion of A genes originally is $p_0 + \frac{1}{2}q_0 = \theta$, say, and under random mating the offspring receives two independent (random) genes from the population: hence $p(AA) = \theta^2$, which is (3). Similarly, $p(Aa) = 2\theta(1-\theta)$ and $p(aa) = (1-\theta)^2$. These are the proportions of the three genotypes in the first generation, so denote them by p_1, q_1, r_1, respectively. The proportion of A genes is $p_1 + \frac{1}{2}q_1$, which a little algebra shows is still θ. Hence, if p_2 is the proportion of the first genotype in the second generation, $p_2 = p_1$, etc., and the proportions of the genotypes remain constant in all subsequent generations at p_1, q_1 and r_1. This is known as Hardy's law.

Example 2. In contrast with random mating consider the same situation with inbreeding, where an individual of one genotype always mates with an individual of the same genotype.

Letting p_n, q_n and r_n be as before for the nth generation, we easily have, corresponding to (2), since

$$p(AA \times AA) = p(AA) = p_n, \quad \text{etc.,}$$

$$p_{n+1} = p_n + \tfrac{1}{4}q_n,$$
$$q_{n+1} = \tfrac{1}{2}q_n,$$
$$r_{n+1} = \tfrac{1}{4}q_n + r_n;$$

or, in matrix notation,

$$\mathbf{p}_{n+1} = \mathbf{A}\mathbf{p}_n, \tag{4}$$

where $\mathbf{p}_n$ is the column vector of elements p_n, q_n and r_n and $\mathbf{A}$ is the matrix

$$\begin{pmatrix} 1 & \frac{1}{4} & 0 \\ 0 & \frac{1}{2} & 0 \\ 0 & \frac{1}{4} & 1 \end{pmatrix}.$$

Consequently

$$\mathbf{p}_n = \mathbf{A}^n\mathbf{p}_0. \tag{5}$$

It is easy to verify (by induction or by summation of a geometric progression) that

$$\left.\begin{aligned} p_n &= p_0 + [\tfrac{1}{2} - (\tfrac{1}{2})^{n+1}]\, q_0, \\ q_n &= (\tfrac{1}{2})^n q_0, \\ r_n &= [\tfrac{1}{2} - (\tfrac{1}{2})^{n+1}]\, q_0 + r_0, \end{aligned}\right\} \tag{6}$$

so that the proportions are not stable after the first generation, as with random mating, though they tend rapidly (because of the factor $(\frac{1}{2})^n$) to the limiting form,

$$p_\infty = p_0 + \tfrac{1}{2}q_0, \quad q_\infty = 0, \quad r_\infty = r_0 + \tfrac{1}{2}q_0,$$

which, as in Hardy's law, will be maintained in subsequent generations, because only pure lines, *AA* or *aa*, are present. In more complex examples an equation of the same matrix type as (5) persists and may be solved by raising the matrix **A** to the appropriate power. Methods for doing this are given in many text-books on algebra; for example, Frazer, Duncan and Collar (1946) or Faddeeva (1959). They all depend on the characteristic roots of the matrix, and the limiting behaviour depends on their values. The columns of the matrix sum to one (since the entries are probabilities) and therefore at least one of its characteristic roots is unity: here unity is a double root. These matrix techniques will be discussed again in connexion with the closely related Markov chains in §4.5.

Example 3. Consider the case where the allele a is recessive so that the two genotypes, AA and Aa, are recognizable as a single phenotype, $\overline{aa}$ say; that is, not aa. The matings allowed in inbreeding will now be $aa \times aa$ and $\overline{aa} \times \overline{aa}$, so that $AA \times Aa$ can occur. The phenotype aa will breed true but even with inbreeding the phenotype $\overline{aa}$ will produce some offspring of the other phenotype aa. We consider how the proportions of these offspring change with successive generations. Considering only matings $\overline{aa} \times \overline{aa}$, the three possible types of mating of the nth generation, with their probabilities, are:

$$\begin{array}{ccc} AA \times AA & AA \times Aa & Aa \times Aa \\ (1-p_n)^2 & 2p_n(1-p_n) & p_n^2, \end{array}$$

where p_n is the proportion of heterozygotes (Aa's) in the nth generation of $\overline{aa}$'s: that is, $p(Aa \mid \overline{aa}) = p_n$. The probability of a heterozygote in the next generation is, again by the generalized addition law,†

$$\begin{aligned} p(Aa) &= p(Aa \mid AA \times Aa)\, p(AA \times Aa) \\ &\quad + p(Aa \mid Aa \times Aa)\, p(Aa \times Aa) \\ &= p_n(1-p_n) + \tfrac{1}{2}p_n^2 = p_n(1-\tfrac{1}{2}p_n). \end{aligned}$$

Similarly the probability of the other phenotype is $p(aa) = \frac{1}{4}p_n^2$. Hence the probability of the phenotype, $\overline{aa}$, is $1-\frac{1}{4}p_n^2 = p(\overline{aa})$. Consequently, in the next generation, by theorem (1.4.2),

$$\begin{aligned} p_{n+1} &= p(Aa \mid \overline{aa}) = p(Aa)/p(\overline{aa}) \\ &= p_n(1-\tfrac{1}{2}p_n)/(1-\tfrac{1}{4}p_n^2) = p_n/(1+\tfrac{1}{2}p_n). \qquad (7) \end{aligned}$$

This equation may be rewritten, $p_{n+1}^{-1} = p_n^{-1} + \frac{1}{2}$,

whence $$p_{n+1}^{-1} = p_0^{-1} + \tfrac{1}{2}(n+1)$$

and $$p_n = p_0/(1+\tfrac{1}{2}np_0). \qquad (8)$$

† All these probabilities are conditional on the mating $\overline{aa} \times \overline{aa}$. This condition, being constant throughout, is omitted in agreement with the convention explained in §1.2.

The proportion p_n of heterozygotes in $\overline{aa}$ tends to zero as n^{-1}, not as fast as the proportion of heterozygotes tended to zero (as 2^{-n}) in the non-recessive case.

Example 4. Continuing with the recessive case, sometimes the phenotype aa corresponds to an abnormality and individuals of the two phenotypes are referred to as normal ($\overline{aa}$) and abnormal (aa). It is then of interest to calculate, given a family tree of the parents, the probability that a child of theirs will be normal. Such a calculation may be of value to cousins contemplating marriage. Consider, as a simple example, the probability that the first child of parents, about whose family history nothing is known except that the parents are normal, be abnormal. In the notation just used this is $p(aa\,|\,\overline{aa}\times\overline{aa})$. To calculate this we use theorem 1.4.7, with A_1 (the notation of the theorem) $= AA\times AA$, $A_2 = AA\times Aa$, $A_3 = Aa\times Aa$, so that $\sum_n A_n = \overline{aa}\times\overline{aa}$. Then

$$p(aa\,|\,\overline{aa}\times\overline{aa}) = p(aa\,|\,Aa\times Aa)\,p(Aa\times Aa)/p(\overline{aa}\times\overline{aa}), \quad (9)$$

the other conditional probabilities in the numerator being zero. Let p be the proportion of a genes in the population and assume random mating. Then by Hardy's law the three genotypes are in proportions $(1-p)^2 : 2p(1-p) : p^2$ and $p(\overline{aa}) = 1-p^2$. Hence, inserting these values in (9), we obtain

$$p(aa\,|\,\overline{aa}\times\overline{aa}) = \tfrac{1}{4}\times\{2p(1-p)\}^2/(1-p^2)^2 = p^2/(1+p)^2. \quad (10)$$

Modifications to this formula will be necessary if the mating is not random, but (9) will still be valid. When more is known about the family tree the computations become more involved since corresponding to the known phenotypic structure more genotypic arrangements are possible. The example illustrates the use of theorem 1.4.7 in deriving a probability conditional on an event (phenotypic structure) which can occur in a number of exclusive ways (genotypic structure).

1.6. Degrees of belief

Mathematical probability (§1.2) has been suggested by properties of events and success ratios of those events (§1.1). We now take a step which widens very significantly the applications of probability theory. We pass from events to propositions, which are collections of events: thus the proposition 'a penny, when released, will fall to the floor' is a statement that all events like 'this penny, when released, will fall to the floor' will occur. And we pass from probabilities of events to probabilities of propositions, and axiomatize these in exactly the same way. Thus to certain pairs, A, B, each of which may either be a proposition or an event, is associated a number $p(A \mid B)$ satisfying the three axioms of §1.2. Such an extension is not useful until we are clear what it is that these probabilities are supposed to represent in the real world; clearly they cannot be success ratios since it is highly artificial to imagine repetitions in each of which a proposition is either true or false. If A is a proposition then $p(A \mid B)$ represents one's belief or strength of conviction, that A is true, given B. It is a *degree of belief* in a proposition. At one extreme, if A is believed to be true, $p(A \mid B) = 1$: at the other extreme, if A is believed to be false, $p(A \mid B) = 0$. Other points in the interval $(0, 1)$ express intermediate beliefs between truth and falsehood. In statistics the word *hypothesis* is used instead of proposition because we are usually interested in a proposition whose truth is in doubt and which is only put forward hypothetically. The principal use of a measure of *degree of belief in a hypothesis* is in describing how our appreciation of a hypothesis changes with increasing evidence relevant to the hypothesis. Let E denote one's accumulated knowledge at some point of time. Let A be some event and, to avoid trivialities, suppose its occurrence or otherwise is not included in E. Let H be some hypothesis. Then $p(H \mid E)$ denotes one's degree of belief in H given E: $p(A \mid HE)$ denotes the probability of the event A given the truth of H and given E: $p(H \mid AE)$ denotes one's degree of belief in H given that A has occurred and given E. Usually E is omitted, since it always occurs after the vertical line, and we write, as before with frequency ideas, $p(H)$ for $p(H \mid E)$, etc. $p(H)$ is called the *prior probability* of H. $p(A \mid H)$, which has been used before as the probability of the event A, given H, is also called the *likelihood of H on A*. $p(H \mid A)$ is called the *posterior probability of H*. The main subject matter of statistics is the study of how data (events) change degrees of belief; from prior, by observation of A, to posterior. They change by Bayes's theorem (1.4.6) since degrees of belief obey the axioms of §1.2. In the form of the corollary to the theorem, if $\{H_n\}$ is a set of hypotheses and A is some event,

$$p(H_n \mid A) \propto p(A \mid H_n)\, p(H_n), \tag{1}$$

for $n = 1, 2, \ldots$, and fixed A. In words this reads 'the posterior probability is proportional to the product of the likelihood and the prior probability'. This result describes how events change beliefs. Two hypotheses H_1, H_2, are independent, given A, if

$$p(H_1 H_2 \mid A) = p(H_1 \mid A)\, p(H_2 \mid A) \tag{2}$$

as before, equation 1.3.1. The justification is also as before: they are independent if the degree of belief in H_1 given H_2 and A is the same as that given A alone. This leads to (2).

Extension of an axiom system

A probability system is a representation of certain aspects of the world, namely success ratios of events, but there is no reason why it should not represent other aspects as well: just as the axioms of length could equally well serve as the axioms of mass. To see this consider rods, that is, one-dimensional objects, denoted by $A, B, \ldots$, which may be combined by placing them end-to-end. Let $A+B$ denote the rod formed by so placing A and B. Then each rod has a length, $l(A)$ say, satisfying

$$l(A) \geqslant 0 \quad \text{and} \quad l(A+B) = l(A)+l(B).$$

But, similarly, these rods have masses, $m(A)$ say, and equally

$$m(A) \geqslant 0 \quad \text{and} \quad m(A+B) = m(A)+m(B).$$

Hence length and mass have the same mathematical properties. This coincidence of mathematical properties happens with frequency limits and degrees of belief and they are both called probabilities. The elements A usually represent events, when dealing with success ratios, and propositions, or hypotheses, when dealing with degrees of belief. Hypotheses are collections of events so it is no great step to pass from one to the other and regard certain sets in sample space as hypotheses. It follows that hypotheses combine in the same way as events: for example, if H_1 and H_2 are two hypotheses, $H_1 H_2$ is the hypothesis which is true iff H_1 and H_2 are both true; H_1 and H_2 are exclusive if they cannot both be true.

The sample space is a space of elementary hypotheses about which, as in the case of elementary events, we do not need to be too specific. As just explained, a hypothesis may itself be considered as a collection of events; and since there is no disadvantage in using a complex sample space (§1.2) we might choose to use elementary events to form the sample space. As a result of these considerations we are able to widen our language by using hypotheses as well as events and 'is true' as well as 'occurs'.

The meaning of a 'degree of belief'

We now have to consider what we mean, in the real world, by $p(A \mid B)$. Consider this where A is a hypothesis H and B is E, our accumulated knowledge of events and hypotheses at some point of time. What is the probability of a hypothesis H given E? What could probability mean in this context? Certainly not a success ratio since indefinite repetitions, in some of which H was true, would be highly artificial. We can obtain a clue to this by considering the way in which words related to probability are used in everyday life. Thus we often hear and use phrases like: 'He will probably get a scholarship...', 'The probable cause of death was...', 'The Tories will probably win...'. In none of these examples are natural repetitions conceivable. Let us analyse the following two statements which eminent scientists are reported to have made:

(*a*) 'It is more probable that a cure for cancer will be found through increasing our knowledge of the mechanism of growth of normal cells, than through experimental treatments applied to cancer patients.'

(*b*) 'I am 90% certain that the neutrons have been produced as a result of a thermonuclear reaction' [and not by some other process].

In (*a*) two hypotheses, H_1 and H_2 say, are compared and one is held to be more probable than the other: in other words H_1

and H_2 are given an ordering (one has more than the other) in terms of probability. In (*b*) the ordering is still present, H_1 is 'by thermonuclear reaction' and H_2 'by other means', but in addition a number is attached to each hypothesis, 90 % to one and 10 % to the other. The second scientist is not merely ordering his hypotheses (like ordering points on a line from left to right) but is attaching numbers to them (like giving the exact positions of the points). The numbers are measures of how strongly he believes in the hypotheses, or degrees of belief in the hypotheses, and can be represented by $p(H_i|E)$, where E is the evidence that the two scientists have.† It is this degree of belief that is represented by the mathematics.

But this raises two questions: what does this 90 % mean? and why should degrees of belief obey the axioms suggested by frequency ideas? One operational interpretation of the 90 % is that the scientist (assuming he had no moral objections to gambling) would be prepared to offer a bet at odds of 9 to 1 on H_1 against H_2; that is, a bet in which he would pay out 9 units if H_2 were true and receive 1 unit if H_1 were true (§1.2). Since he is not completely certain about H_1 he cannot be sure of receiving the one unit, and might therefore assess it as worth x units ($x < 1$). That is, a certainty of x units is as good as this chance of receiving 1 unit. Similarly, although he does not think H_2 is true he might lose his 9 units: and this loss might be judged equivalent to a certain loss of $9y$ units ($y < 1$). Presumably he does not expect to be the loser by the offer and therefore $x \geqslant 9y$ or $x/y \geqslant 9$. Let us now make the assumption that the values x and y do not depend on the units used. For example, that a loss of 10 units if H_2 were true would be judged equivalent to a certain loss of $10y$ units. This assumption is reasonable provided the units are not changed drastically, as we have no wish to do. Now the scientist would obviously also be prepared to offer 8 to 1, and this is confirmed since $x/y > 8$. But he might, if pressed, offer 10 to 1. This he would do if $x/y \geqslant 10$, in virtue of the assumption just made. All his statement, quoted

† Notice how often the conditioning event is omitted, as in the frequency cases previously encountered. In (*b*) it was very relevant because later experiments suggested that H_1 was false.

above, has told us is that $x/y \geqslant 9$. By putting other suggestions to him we can determine the exact value of x/y. Thus, if 9 to 1 on H_1 against H_2 is the highest odds he will offer then $x/y = 9$. This upper limit of odds is called the *critical* limit and we denote it by b: so $x/y = b$. As only the ratio of x to y matters we conventionally suppose them to add to one, when $x/(1-x) = b$ or $x = b/(1+b)$. This value of x is *our degree of belief* in H_1 and has the properties of probability, and which we term the probability of H_1. Similarly y is the probability of H_2. The empirical content of the former type of probability was a frequency ratio. The empirical content of this type is in terms of the critical odds for a bet. A bet at the critical odds is said to be *fair*. The reason for this is that, in our case, the scientist would not offer a bet at higher odds because he thinks it unfair to him, and if he made it at lower odds he would stand to win, which would be unfair to his opponent. In language to be developed later (§2.1) a fair bet has zero expectation. The 90 % quoted by the scientist was probably his assessment of a fair bet, since the statement, unlike a bookmaker's, would not be made with expectation of gain.

Justification of the axioms

The answer to the second question at the beginning of the last paragraph is more involved and the next two paragraphs may be omitted by any reader who is prepared to take the axioms on trust. The justification uses the interpretation of degrees of belief in terms of betting and the following easily acceptable basis for argument: if two bets are both fair then so is the bet in which the first bet is taken with probability α and the second with probability $(1-\alpha)$. The probability here is a frequency one: for example with $\alpha = 1/3$, a fair die could be rolled; if it shows a 1 or a 2 the first bet will be taken, otherwise the second will be taken. The bets are said to be *mixed* in the ratio $\alpha:1-\alpha$. The first axiom (§1.2) is immediate since any degree of belief must lie between 0 and 1 (it is $b/(b+1)$): and if H is given, H is certainly true, when any odds will be offered, so $b \to \infty$ and the degree of belief is 1. To justify the second axiom let H_1 and H_2 be exclusive hypotheses, given E, with critical bets of b_1 to 1 and

b_2 to 1. Since these are fair a mixture in the ratio $\alpha : 1-\alpha$ is fair. Table 1.1 may help to clarify the position; the entries are the values of a bet (a row) given the truth of a hypothesis (a column). Since H_1 and H_2 are exclusive the only possibilities are H_1, H_2 or neither H_1 nor H_2, that is $\bar{H}_1 \bar{H}_2$. The values for the mixture are given in the third row and for H_1 and H_2 they are equal if

$$\alpha - (1-\alpha) b_2 = -\alpha b_1 + (1-\alpha),$$

or

$$\alpha = \alpha_0 = (1+b_2)/(2+b_1+b_2).$$

With $\alpha = \alpha_0$ the values are given in the last row. Since the values for H_1 and H_2 are equal, and therefore it does not matter which is true, this α_0 mixture is equivalent to a bet on

$$H = H_1 + H_2$$

and the odds are $(b_1 + b_2 + 2b_1 b_2)$ to $(1 - b_1 b_2)$,

or

$$b = (b_1 + b_2 + 2b_1 b_2)/(1 - b_1 b_2) \text{ to } 1.$$

Hence the degree of belief in H is

$$\frac{b}{1+b} = \frac{b_1 + b_2 + 2b_1 b_2}{1 + b_1 + b_2 + b_1 b_2} = \frac{b_1}{1+b_1} + \frac{b_2}{1+b_2},$$

the sum of the degrees of belief for H_1 and H_2. The extension to several exclusive hypotheses is immediate and it is conventionally used for an enumerable infinity.

TABLE 1.1

True hypothesis ...	H_1	H_2	$\bar{H}_1 \bar{H}_2$
Bet on H_1	$+1$	$-b_1$	$-b_1$
Bet on H_2	$-b_2$	$+1$	$-b_2$
Mixture 'α'	$\alpha-(1-\alpha)b_2$	$-\alpha b_1+(1-\alpha)$	$-\alpha b_1-(1-\alpha)b_2$
Mixture 'α_0'	$\frac{1-b_1 b_2}{2+b_1+b_2}$	$\frac{1-b_1 b_2}{2+b_1+b_2}$	$-\frac{b_1+b_2+2b_1 b_2}{2+b_1+b_2}$

To justify the third axiom again consider two hypotheses H_1 and H_2, given E, which are not exclusive. Let the critical bet for H_1 be b_1 to 1, and for H_2, given H_1 (and E), be b_2 to 1. In the latter bet b_2 units are paid out if H_1 is true but H_2 not, and 1 unit is received if H_1 and H_2 are both true, otherwise there is no payment. In other words, the bet on H_2 is only operative if H_1

obtains. The values are given in table 1.2, together with the values for a mixture in the ratio $\alpha:1-\alpha$. The values for this latter bet are equal when $H_1\bar{H}_2$ or $\bar{H}_1$ are true if

$$\alpha-(1-\alpha)b_2 = -\alpha b_1 \quad \text{or} \quad \alpha = \alpha_0 = b_2/(1+b_1+b_2).$$

With $\alpha = \alpha_0$ the values are given in the last row. This is a bet in favour of H_1H_2 at odds of $b = b_1b_2/(1+b_1+b_2)$ to 1. Hence the degree of belief in H_1H_2, given E, is $b/(1+b) = p(H_1H_2|E)$, and

$$p(H_1H_2|E) = b_1b_2/(1+b_1)(1+b_2) = p(H_1|E)\,p(H_2|H_1E),$$

which is the third axiom. A more complete justification for the axioms for degree of belief have been given by Savage (1954). Savage also gives reasons for thinking that everyone must be able to assess $p(H|E)$, that is, that it always exists, a question we now proceed to discuss.

TABLE 1.2

True hypothesis ...	H_1H_2	$H_1\bar{H}_2$	$\bar{H}_1$
Bet on H_1	$+1$	$+1$	$-b_1$
Bet on H_2 given H_1	$+1$	$-b_2$	0
Mixture 'α'	$+1$	$\alpha-(1-\alpha)b_2$	$-\alpha b_1$
Mixture 'α_0'	$+1$	$-\dfrac{b_1b_2}{1+b_1+b_2}$	$-\dfrac{b_1b_2}{1+b_1+b_2}$

Evidence for the numerical assessment of a belief

It has been shown that a degree of belief, if measured numerically, will satisfy the same axioms as are suggested by frequency considerations. But the question remains: can a degree of belief be measured numerically? No one would dispute that we all have beliefs, some more strongly held than others, but many people object to any attempt to describe them in numerical terms. We now give an outline of the method that is used to establish the existence of *numerical* degrees of belief.

The method is to consider the way in which any person must act, to impose on his actions sensible restrictions that almost all people would accept as rational, and to demand that his actions be consistent. From axioms concerning actions it is possible to deduce, by the usual methods of mathematics, that the only way of acting that is rational and consistent must involve a numerical statement of relevant degrees of belief, or must be equivalent to such a statement. This method was used by von Neumann and Morgenstern (1947) to consider utility (a topic that is briefly discussed in §5.6) and extended by Savage (1954) to cover degrees of belief, or what he calls *personal probability*.

The axioms that Savage considers concern two sets of elements that he calls the 'states' and the 'consequences'. The states are the possible hypotheses concerning the topic under discussion. For example, in the situation (*b*) above the states are the various possible ways in which the observed neutrons could have been produced. One of these is by a thermonuclear reaction. The probabilities concern these states. The consequences are what will ensue if actions are taken on the basis of the beliefs. Again in the example from nuclear physics, one can imagine actions, costing considerable sums of money, being taken on the basis of beliefs (or opinions) about the process which produced the neutrons. An action, or a decision, can be thought of as a function which defines, for each state, a consequence. Thus the action of building a larger machine will result in the production of more neutrons and considerable benefits if the thermonuclear reaction was the cause; but it will be largely wasted if it was not the cause:—that is if some other state obtained.

It is then possible to develop axioms concerning these elements, the states and the consequences, and the acts which relate them. The first axiom is that all acts can be ordered: that is, given any two, one is preferred to the other (or they are equally preferred) and if a_1 is preferred to a_2, and a_2 to a_3, then a_1 is preferred to a_3. This axiom is a severe assumption but one that is difficult to attack. For it is always conceivable that you may have to decide between two given acts and, if they are the only ones available to you, you will act in one of those two ways. That choice of action is itself an ordering; namely that the rejected action is not preferred to the chosen one. Thus any person deserving to be called rational must order the acts available to him.

A second axiom is that if action a_1 is preferred to action a_2 when the state of the world is s_1, and a_1 is still preferred to a_2

when the state is s_2, then a_1 is preferred to a_2 even if it is not known whether the state is s_1 or s_2 (that is, irrespective of beliefs about s_1 or s_2). For example, if it is better to build a larger machine than not to, when the neutrons were produced by a thermonuclear reaction, and also better if they were produced differently, then it is better to build the machine in any state of mind about how they were produced. This is essentially an axiom of consistency on the part of the person taking the action.

These are the basic axioms. By adding others of a less fundamental nature—just as the first axiom in §1.2 is not so basic as either of the others—it is possible to demonstrate the existence of numbers attached to the states, which satisfy our axioms of probability, and which therefore deserve to be called degrees of belief probabilities. It is also possible to demonstrate the existence of a utility function (for a brief discussion of this see §5.6).

The attitude we adopt in this book is that these axioms are reasonable, that the deductions from them described in detail by Savage are correct, and that therefore we may use probabilities in the sense of degrees of belief. It must be pointed out, however, that in adopting this attitude, and developing statistics in chapters 5–8 on this basis, we are pursuing an unorthodox approach. Most statisticians adopt a different attitude in which only frequency probabilities are admitted. The present approach seems preferable to the author. The point is discussed more fully at the end of §5.6. The student who learns his statistics from this book need not fear that he will learn different results. The methods developed are the same, in all essentials, for the two schools of thought. We indicate the few points where they differ. The distinction between the two schools is irrelevant in chapters 2–4.

Likelihood

The word likelihood was introduced by Fisher, who was responsible for many of the important ideas in modern statistics. It may seem unusual to have two different words for the same quantity. The reason is this: in some problems, usually deductive ones, one is concerned with the probabilities of different events

$A_1, A_2, \ldots$ all given the same hypothesis H, so that in $p(A_i \mid H)$, A_i is the variable. Whereas in other problems, usually inductive ones, an event, A, has occurred and (usually in order to apply Bayes's theorem) one is interested in the probability of A given different hypotheses, $H_1, H_2, \ldots$ (compare the insulin example, below), so that in $p(A \mid H_i)$, H_i is the variable. Therefore, depending on which is the variable, or the argument of the probability function, the terms probability ($p(A_i \mid H)$) or likelihood ($p(A \mid H_i)$) are used. One speaks of the probability of an event or the likelihood of a hypothesis. For example, one difference between probability and likelihood is that if $\{A_i\}$ are exclusive and exhaustive the probabilities sum to one, whereas if $\{H_i\}$ are exclusive and exhaustive the likelihoods do not usually sum to one.

Relation between the two types of probability

In this book we have chosen to develop frequency ideas first and then pass to degrees of belief: we do this because the former are the easier to understand. It is possible, and perhaps even desirable, to proceed the other way round; starting from a notion of probability developed from beliefs and deducing frequency ideas. This has been done most successfully by de Finetti (1937); an account of some of his work in English is given by Savage (1954).

We have, we hope, thus justified the use of two types of probability. These can occur together in an equation. Furthermore both interpretations are sometimes possible, and when this is so the two values agree. Thus, suppose we have a newly minted coin, one that is known to be just as likely to fall heads or tails. Then our degree of belief that it will fall heads in a single toss is 1/2, as is the frequency probability. Of course, if we have a bent coin the frequency probability may be p, say, and our degree of belief may be $p' \neq p$. But that is because the conditioning event is different in the two cases. The frequency probability can only be stated if p is known, when $p(\text{heads} \mid p) = p$. The degree of belief is the probability of heads without knowledge of p. If we tossed the bent coin a large number of times we should learn the value of p and the degree of belief in heads would move towards p (see §7.2).

There is some disagreement among writers about the objectivity of $p(H \mid E)$. Some maintain that it is a subjective value and that two people will not necessarily assign the same number to H, given E. The view taken in this book is that if two people differ in their probabilities of H, where H is a scientific question and not, for example, a question of taste, it is because they have different evidence; in other words, one is considering $p(H \mid E)$ and the other $p(H \mid E')$. If they were to pool their knowledge then they should agree on $p(H \mid EE')$. It is easy to make the mistake of thinking the probabilities are different, when it is really the evidence that is different, because, as repeatedly mentioned, the evidence is often omitted in the statement of a probability. Of course, it is sometimes difficult to pool the knowledge of different people: but it is important to do so in order to reach a common assessment of probability, and the need to do this is an important reason for improving scientific communication. Attempts have been made to develop statistical theories which use only probabilities of events, which are therefore objective, but none have been entirely successful. Reasons for rejecting them will be given later (§5.6).

Scientific use of Bayes's theorem

Although Bayes's theorem is basic in understanding how we accumulate knowledge, it must be noted that in a well-designed experiment the result it gives is often trivial. Suppose that there are only two hypotheses, H_1 and H_2, under consideration and that there exists an experiment in which an event A has, under the hypotheses, the probabilities $p(A \mid H_1) = 1$ and $p(A \mid H_2) = 0$; whence it follows that $p(\bar{A} \mid H_1) = 0$ and $p(\bar{A} \mid H_2) = 1$. Then, provided neither of the prior probabilities $p(H_1)$ nor $p(H_2)$ is zero, whether A or $\bar{A}$ occurs one of the posterior probabilities will be zero and the other non-zero. The experiment is conclusive in favour of one of H_1 or H_2. (If A, H_1 is true; if $\bar{A}$, H_2 is true.) This then is an excellent experiment for distinguishing between the rival hypotheses and the result is obvious without an appeal to the theorem. It is the method used by scientists throughout scientific history. An excellent example is the experiment on the curvature of light rays during an eclipse which decided the issue between the special theory of relativity and Newton's laws. It often happens, however, that it is not possible to design such an unequivocal experiment. For example, the measurement of potency, or the assay, of insulin, is carried out with animals. Some animals are affected by a given dose and some are not, and the frequency concept applies to give the probability of an animal being affected, which can be used as a measure of the potency. Trials on a number of animals with a fixed type of insulin form random trials with constant probability of success (= affected animal), compare §1.3. Suppose that the required potency of standard insulin is such that the probability is 1/2. If, in an experiment on 20 animals, 15 are affected does this mean that the insulin is too potent? 15 could be affected even if p was 1/2, just as one could get 15 heads in 20 tosses of a fair coin, but 15 is more probable when $p = 3/4$. So this experiment is not completely successful in deciding between these two values of p, it suggests, but nothing stronger, that p is nearer 3/4 and 1/2. Nevertheless, apart from using more animals or other refinements such as measuring the effect, it is the best available experiment. Such experiments will be discussed from chapter 5 onwards, but particularly in §7.2.

Suggestions for further reading

There are many books which discuss the foundations of probability and, in particular, the relationship to the frequency concept and the concept of a degree of belief. A modern reference which has had a substantial influence on the present author is Savage (1954). Savage's work is derived from that of Ramsey (1931); de Finetti, references to whose work are given in Savage, a convenient article is (1937); and the axiomatic approach of von Neumann in the appendix to (1947). Other important books are those by Keynes (1921), Reichenbach (1949), Braithwaite (1953) and Carnap (1950). Two historical references, easily available in modern printings, are Laplace (1951) and Bayes (1958). Few of these books attempt to go beyond a discussion of the foundations and apply the results: notable exceptions are Jeffreys (1961) and Good (1950), both of which include substantial developments of the calculus.

Two books that discuss the mathematical calculus of probabilities are Feller (1957) and Gnedenko (1962). Both authors adopt a frequency view-point and dismiss any other attitude toward probability, but since their primary concern is with the mathematics this scarcely matters. The first book to present the modern axiomatic theory is still worth reading, Kolmogorov (1956). Modern probability theory is based on concepts of measure and Lebesgue–Stieltje integration. These topics are discussed by Munroe (1953), Halmos (1950) and Pitt (1963).

Some people enjoy the combinatorial side of probability which we have not discussed. Essentially this deals with the special case where all the elementary events are equally likely and the problems concern the computation of probabilities of other events, often involving permutations and combinations. A delightful elementary book in this field, which anyone who enjoys puzzles would appreciate, is Whitworth (1901). Two modern, more sophisticated, books are by David and Barton (1962) and Riordan (1958). A classic in this branch of the subject is MacMahon (1916).

Exercises

1. Roll a die, toss a coin, spin a roulette wheel or use some other random device with an appropriate event for success, and plot the success ratio against the number of trials as the latter increases. Notice the decrease in oscillation of the ratio as the trials grow in number.

(This exercise is best done in a class in order to combine all the results into a single series. If a suitable random device is inconvenient tables of random sampling numbers (§3.5) may be used instead.)

2. At the beginning of each period of 1 minute duration there is a one-third chance of a customer arriving at a counter to be served. If the server is free he serves the customer for 2 minutes. If the server is not free the customer joins a queue and is served, again for a period of 2 minutes, as soon as the customer immediately in front of him has been served. By rolling a die and assuming a customer arrives if a 1 or 2 is exposed, follow through such a process for at least 30 periods and record:

(*a*) the proportion of customers who did not have to wait;
(*b*) the proportion of time the server was free;
(*c*) the maximum time any customer had to wait;
(*d*) the maximum size of queue formed.

[Imitation of a process using random numbers is called *simulation*.]

3. If you had only a fair coin available how could you have produced an event of probability one-third for the above example?

4. Simulate (see exercise 2) random mating of two alleles A and a with three distinct phenotypes AA, Aa, aa; in which the allele A is twice as frequent as a in the population. Suppose that each mating produces two independent offspring (all parents have two children).

5. Seven tickets are numbered consecutively from 1 to 7. Two of them are selected in order without replacement. Enumerate the sample space of possible results of the selection and identify the following events:

A, the numbers on the two tickets add up to 9;
B, both tickets have prime numbers (count 1 as a prime);
C, the numbers on the two tickets differ by 3.
Identify also the events AB, BC, $A+B$.

If each elementary event has probability 1/42 (the event 1,7 being counted as distinct from 7,1), find $p(A|B)$, $p(B|C)$, $p(C|A)$.

6. Each member of a group of persons has his height and weight recorded. Each point of the sample space (elementary event) may be represented as a point on the paper with height as abscissa and weight as ordinate. Identify the following events in a Venn diagram:

A, height over 5 ft. 11 in., weight over 12 stone;
B, weight over 2 lb per inch of height;
C, weight between 9 and 11 stone.

Identify the events AB and $\overline{BC}$.

7. A needle of length l is thrown onto a level table marked with parallel lines at distance $h\,(>l)$ apart. Prove, under reasonable assumptions which should be stated, that if the needle makes an angle θ with the lines the probability that it will cross a line is $l\sin\theta/h$.

8. Each point in a plane has co-ordinates with respect to a pair of rectangular axes. A circular disc of radius $r < \frac{1}{2}$ is thrown on to the plane. What is the probability that it will cover some point whose co-ordinates are both integers?

9. A circle of diameter $\frac{4}{3}$ in. is placed with its centre at a randomly determined point in a square of side 1 in. Prove that, if $P(r)$ is the probability that exactly r corners of the square lie in the ring,

$$P(2) = \frac{16}{9}\cos^{-1}\frac{3}{4}-\frac{\sqrt{7}}{3},$$

and

$$P(1) = \frac{4\pi}{9}-2P(2).$$

(Aberdeen Dipl.).

10. Comment on the dependence and independence of height, weight and colour in a population consisting of 25 % short-thin-coloured, 25 % short-fat-white, 25 % tall-thin-white and 25 % tall-fat-coloured persons.

11. A finite discrete sample space consists of the four points denoted

$$(110),\quad (101),\quad (011),\quad (000),$$

and each has a probability 1/4. The event A_i ($i = 1, 2, 3$) occurs if there is a 1 in the ith place: thus, A_1 contains the two points (110) and (101). Show that A_i and A_j $(i \neq j)$ are independent but that A_1, A_2 and A_3 are not independent.

Construct a similar example to show that three events A, B, C can satisfy

$$p(ABC) = p(A)\, p(B)\, p(C)$$

without any pair of them being independent.

12. A box contains n balls, r of which are white and $n-r$ black. Two balls are drawn at random: (i) with replacement, (ii) without replacement. Calculate the probabilities that both balls be white, one ball be white and one black, in the two cases. Show that if $n \to \infty$, with $r/n = p$ fixed, the limits are the same in the two cases (i) and (ii).

13. Show that if A is independent of each B_i ($i = 1, 2, \ldots, n$) and the B_i are exclusive, then A is independent of ΣB_i. Show, by an example with $n = 2$, that this result is not necessarily true when the B_i are not exclusive.

14. An urn contains M balls of which $W(<M)$ are white. $n(<M)$ balls are drawn and laid aside (not replaced in the urn), their colour unnoted. Another ball is drawn. What is the probability it will be white?

15. Colour-blindness appears in 1 % of the people in a certain population. How large must a random sample (with replacements) be if the probability of its containing a colour-blind person is to be 0·95 or more?

16. An event A is said to *favour* an event B if $p(B|A) > p(B)$. If A favours B, and B favours C, does it follow that A favours C? If so, prove the result: if not, give an example where the result is false.

17. Circular discs of radius r are thrown at random on to a plane circular table of radius R which is surrounded by a border of uniform width r lying in the same plane as the table. If the discs are thrown independently and at random, and N stay on the table, show that the probability that a fixed point on the table, but not on the border, will be covered is

$$\left\{1-\left(1-\frac{r^2}{(R+r)^2}\right)^N\right\}.$$

(Camb. Dipl.)

18. Prove that

$$p(A+B+\ \ldots\ +Z) \leqslant p(A)+p(B)+\ \ldots\ +p(Z).$$

Express $p(A+B+C)$ in terms of $p(A)$, $p(B)$, $p(C)$, $p(BC)$, $p(AC)$, $p(AB)$, $p(ABC)$.

Prove that $p(AB \ldots Z) \geqslant 1-p(\bar{A})-\ \ldots\ -p(\bar{Z})$.

19. A, B are two events. In terms of $p(A)$, $p(B)$, $p(AB)$ for $k = 0, 1, 2$ express:

(i) p (exactly k of A, B occur);
(ii) p (at least k of A, B occur);
(iii) p (at most k of A, B occur).

20. A certain university course contains two examinations, one of which, known as 'Prelim', is taken at the end of the first year and the other, known as 'Part I' at the end of the second year. Results in Prelim are classified as 1st, 2nd, 3rd or fail: the probabilities of a randomly selected candidate being classed in these groups are respectively, p_1, p_2, p_3 and p_4. Candidates for Part I are classified as 1st, 2nd (upper division), 2nd (lower division), 3rd or fail. A randomly selected student who has been placed in group i in Prelim has chances $p_{i1}, p_{i2}, p_{i3}, p_{i4}$ and p_{i5} of being placed in the 5 groups possible in Part I. (Assume that a candidate who fails in Prelim does not take Part I.) Write down the probabilities of the following events:

(i) a college with 24 randomly selected candidates having less than 3 failures in Prelim;

(ii) a randomly selected candidate obtaining a 1st class in Part I;

(iii) a candidate getting a better class in Part I than in Prelim;

(iv) a candidate failing at some stage;

(v) in a group of six friends one obtaining a 1st class both in Prelim and Part I, two obtaining 1st class in Prelim and 2nd class in Part I and three obtaining 2nd class both in Prelim and Part I. (Camb. N.S.)

21. An r-partition of n is a collection without regard to order of r integers ($\geqslant 1$) whose sum is n. Galileo was given, and solved, the following problem. There are six 3-partitions with no integer above 6 of 10 and of 12 yet when 3 six-sided fair dice are thrown simultaneously the total 10 appears more frequently than 12. Why?

22. A man tosses a coin either until he gets two heads in succession or until he has three tails (not necessarily in succession). Enumerate the sample space. If the probability of a head is 1/2 and is uninfluenced by the other tosses, calculate the probabilities attached to each elementary event. What is the probability that he will toss the coin more than three times?

23. Suppose that in answering a question in a multiple choice test, an examinee either knows the answer, with probability p, or he guesses, with probability $1-p$. Assume that the probability of answering a question correctly is unity for an examinee who knows the answer and $1/m$ for the examinee who guesses, where m is the number of multiple choice alternatives. Show that the probability that an examinee knew the answer to a question, given that he has correctly answered it, is

$$\frac{mp}{1+(m-1)p}.$$

24. A sequence of 4 signals, 'on' or 'off' is produced in the following way. The probability that the first is 'on' is 1/2 (and therefore 'off' is 1/2). The probability that the rth signal is the same as the $(r-1)$th is 0·9 (and therefore that it is different is 0·1) for $r = 2, 3, 4$. Calculate the probabilities of s 'on' and $4-s$ 'off' for $0 \leqslant s \leqslant 4$. Compare these values with the same probabilities calculated on the assumption that each signal is equally likely to be 'on' or 'off' irrespective of the other signal.

25. There are 6 men in a room, 2 pairs of brothers and 2 unrelated men. The probability that any one of them has blood-group X is 1/4. The probability that if one brother has blood-group X, the other brother has also X is 3/4, otherwise the blood-groups are independent. Calculate the probability that exactly 3 men in the room have blood-group X.

26. Two players, A and B, play the following game with a fair die. A throws the die and then has a second set of throws equal in number to that shown on the die the first time it was thrown. The total, s, on the second set of throws is recorded. If $s > 12$, B gives A one unit. If $s = 12$ there is no payment. If $s < 12$, A gives B one unit. What is the probability that $s = 12$? Is it better to be A or B? (This problem is due to James Bernoulli.)

27. Two players X and Y play the following game with a perfect six-sided die with sides labelled 1, 2, ..., 6. At each round, the players shake the die (X shaking first), and the first to obtain a six wins. If in any round a player shakes a 1, then he has one more shake in the same round (but only one more, whatever the outcome).

What is the probability that the game ends in either the first or the second round? What is X's chance of winning the whole game?

(London Ancil.)

28. Two players take turns in drawing a ball out of an urn containing three white and five black balls. The player who begins requires a white ball, the other a black, he who first obtains a ball of his required colour winning the game. Find the relative chances of winning:

(i) if each ball is replaced before the next draw;
(ii) if no replacements are made.

29. The following are the rules for the game of 'Craps'. A player has two dice which are thrown together. If the total is 7 or 11 he wins (a 'natural'): if 2, 3 or 12 he loses (a 'crap'). If neither of these results, he continues throwing until either he repeats his original score, when he wins, or he throws a 7, when he loses. Show that the probability of losing is 251/495. (Assume that the dice are not 'loaded'.)

30. Assume that the genotype frequencies in a population are $P = p^2$, $H = 2pq$, $Q = q^2$. Given that an offspring, O, is of genotype Aa, show that the probability that a sibling (another offspring of O's parents) is of the same genotype is $\frac{1}{2}(1+pq)$.

31. A male rat is either doubly dominant (AA) or heterozygous (Aa) and, owing to Mendelian properties, the probability of either being true is 1/2. The male rat is mated with a doubly recessive (aa) female. If the male rat is doubly dominant, the offspring will exhibit the dominant characteristic; if heterozygous, half of the offspring will exhibit the dominant characteristic and the other half the recessive characteristic. Suppose of three offspring all exhibit the dominant characteristic. What is the probability that the male is doubly dominant?

32. In each individual there is a pair of genes, each of which may be of type X or of type x. Individuals with Xx or xX are termed heterozygotes; those with xx are abnormal; both heterozygotes and those with XX are normal. The proportion of abnormal individuals of either sex in a population is p^2 and of heterozygotes is $2p(1-p)$, where $0 < p < 1$. Each of the parents of a child transmits one of its own genes to the child; if a parent is a heterozygote, the probability that it transmits the gene of type X is $1/2$. Mating in the population is to be assumed to be random. Show that among normal children of normal parents, the expected proportion of heterozygotes is $2p/(1+2p)$.

James, a normal child of normal parents, marries a heterozygote and they have n children, all normal. Find, in the light of this information, the probability that James is a heterozygote and the probability that his first grandchild will be abnormal. (Camb. Tripos.)

33. A certain phenomenon is either present or not. An apparatus has a probability $p_1 > \frac{1}{2}$ of detecting the phenomenon (by giving a positive response) if it is present and a probability $p_2 < \frac{1}{2}$ of giving a positive response even if it is absent. If the phenomenon is equally likely to be present or absent and three independent uses of the apparatus give two positive responses and one negative what is the probability that the phenomenon is present? What is the probability that a fourth test will yield a positive response?

34. If $H_1, H_2, \ldots, H_n$ are exclusive and exhaustive hypotheses with degrees of belief $\pi_1, \pi_2, \ldots, \pi_n$ then

$$I = \sum_{i=1}^{n} \pi_i \ln \pi_i$$

is called the *information* about the H_i. Show that I is least when $\pi_i = n^{-1}$ for all n.

Apply this to exercise 33 and find the information available about the presence of the phenomenon before and after the three responses had been obtained.

35. If a person is suspected of having an undesirably high alcohol content in his blood a rapid test is carried out. This test has only 75% chance of being correct: i.e. of giving a positive response when the alcohol level is high, or of giving a negative response when it is low. If the rapid test gives a positive response the subject is taken to a laboratory and given a second test. If the alcohol content was high at the time of the original test, this test has 90% chance of detecting it (the delay in getting to the laboratory accounts for the 10% errors). If the content was low the second test never gives a false result. These are independent of the first test. If a person is initially classed as a suspect when the prior probability of his having a high alcohol content is 20% answer the following questions:

(i) What proportion of suspects will have a second test which does not detect alcohol?

(ii) What is the posterior probability that such a person had a high alcohol content?

(iii) What proportion of suspects will not have a second test?

(iv) If a person is not tested in the laboratory, what is the posterior probability that he had a high alcohol content?

36. In a bolt factory, machines A, B and C manufacture 25, 35 and 40% of the total respectively. Of their output, 5, 4 and 2% are defective bolts, respectively. A bolt is drawn at random from the production and is found defective. What are the probabilities that it was manufactured by machines A, B and C?

37. A newspaper reported that 'the odds are two to one against the Government bringing in the bill....If it does bring it in, reliable sources suggest that the odds are about even that it will get through.' What is the probability that the bill will become law?

38. A patient thinks he may have cancer and consults his doctor, A, who after examination declares he has not. The patient feels his doctor is over-cautious about diagnosing cancer so he consults a second doctor, B, who declares that he has cancer. How is the patient's belief that he has cancer affected by these two medical opinions?

Show that, if doctor A diagnoses cancer in only 60% of those patients who have it, and never in the case of those who do not: and if doctor B diagnoses cancer in 80% of those patients who have it and in 10% of those who do not, then the patient's odds against not having cancer are multiplied by 16/5 as a result of the two opinions (which are supposed given independently).

39. $A_1, A_2, \ldots, A_n$ are events which could result from an experiment. Show that the probability that at least one of these events occurs is given by

$$P_1 = \sum_i p(A_i) - \sum_{i<j}\sum p(A_i A_j) + \sum_{i<j<k}\sum\sum p(A_i A_j A_k) - \ldots + (-1)^{n-1} p(A_1 A_2 \ldots A_n),$$

where, for example, $p(A_i A_j)$ is the probability that both A_i and A_j occur.

Sixteen married couples are the only people involved in an accident from which only six persons survive. Calculate the probability that there is at least one married couple among the survivors. (Leicester.)

2

PROBABILITY DISTRIBUTIONS—ONE VARIABLE

In the first chapter the discussion concerned the occurrence or non-occurrence of an event (or the truth or falsity of a hypothesis). Observations are often more detailed than merely looking to see if an event has occurred or not, and in this chapter the concept of probability is extended to the case where the observation is a single real number.

2.1. The discrete case

Consider a random sequence of n trials with constant probability, p, of success (§1.3). We evaluate the probability, p_r, of r successes (and therefore $(n-r)$ failures) in the n trials. The r successes and $(n-r)$ failures can take place in several orders. Consider any particular order, for example r successes followed by $(n-r)$ failures. Since the trials are random, that is the events of success or failure are independent, the multiplication law (equation 1.3.2) applies and the probability of obtaining r successes followed by $(n-r)$ failures is $p^r q^{n-r}$, where $q = 1-p$, the probability of a failure. This probability is the same for any other order in which the r successes occurred. But the different orders are certainly exclusive, and there are $\binom{n}{r}$ of them,† so that the addition law (axiom 2, §1.2) applies and the required probability is given by

$$p_r = \binom{n}{r} p^r q^{n-r}. \tag{1}$$

The situation just considered is a particular case of the following. Associated with each elementary event a in the sample space is an integer which may be positive or negative or

† $\binom{n}{r}$ denotes the number of combinations of n things taken r at a time. It is equal to $n!/r!(n-r)!$ and is alternatively written ${}^nC_r$.

zero; A_r is the set of a for which the associated integer is r and $p_r = p(A_r \mid \mathrm{A}) = p(A_r)$, say. The events $\{A_r\}$ are exclusive and exhaustive, so the $\{p_r\}$ satisfy:

$$p_r \geqslant 0, \quad \sum_r p_r = 1. \tag{2}$$

A sequence $\{p_r\}$ satisfying (2) is said to be a *probability density*. The integer associated with each a is called a *random variable*, and $\{p_r\}$ is the probability density of the random variable. Probability density will usually be abbreviated to *density*. The random variable is said to have a *distribution* (of probability), and one way of describing the distribution is by its density. The special distribution whose density is given by (1) is called the *binomial* distribution: n is termed the *index* and p the *parameter* of the distribution. A random variable with a binomial distribution is called a *binomial variable* and we often say a random variable is $B(n, p)$, meaning that it is binomial with index n and parameter p.

If $\{p_r\}$ is a density (of a random variable)

$$P_r = \sum_{s \leqslant r} p_s$$

is called a *distribution function* (of a random variable) and provides another way of describing a distribution. Clearly, $\sum_{s \leqslant r} A_s$ is the event that the random variable is less than or equal to r. Hence by the addition law, P_r is the probability that the random variable is less than or equal to r. We also have that

$$P_r \leqslant P_s \quad \text{if} \quad r \leqslant s, \tag{3}$$

and since $\sum_r p_r$ is a convergent series of non-negative terms,

$$\lim_{r \to -\infty} P_r = 0, \quad \lim_{r \to +\infty} P_r = 1. \tag{4}$$

Any function satisfying (3) and (4) is said to be a distribution function.

If the series
$$\Sigma r p_r \tag{5}$$
converges absolutely (that is if $\Sigma |r| p_r$ converges) its sum is called the *mean* of the distribution, and is usually denoted by μ, or the *expectation* of the random variable with this distribution.

In the latter context it is usual to write (5) as $\mathscr{E}(r)$. If $f(r)$ is a function of the random variable,

$$\Sigma f(r)\, p_r \tag{6}$$

is, if absolutely convergent, called the *expectation* of $f(r)$, and (6) is written $\mathscr{E}[f(r)]$. The following theorems are immediate from the usual properties of absolutely convergent series, where the terms may be summed in any order.

Theorem 1. *If $f(r)$, $g(r)$ are two functions of a random variable whose expectations exist, then*

$$\mathscr{E}[f(r)+g(r)] = \mathscr{E}[f(r)]+\mathscr{E}[g(r)]. \tag{7}$$

Theorem 2. *If the random variable is non-negative then its expectation is*

$$\sum_{r=0}^{\infty}(1-P_r). \tag{8}$$

For $$\sum_{r=0}^{\infty}(1-P_r) = \sum_{r=0}^{\infty}\sum_{s>r} p_s = \sum_{s=0}^{\infty}\sum_{r=0}^{s-1} p_s = \sum_{s=0}^{\infty} s p_s.$$

Binomial examples

There are many applications of the binomial distribution and we consider only a few. In factory inspection a random sample (§1.3) is taken from a lot and the items in the sample are each tested to see if they are satisfactory (success) or defective. If the sample is small compared with the lot size (see §1.3 again) the distribution of the number, r, of defectives is, because of the way the sample has been taken, binomial with index n and parameter p, the proportion of defectives in the lot. The distribution of r tells one how often, in samples of size n for a given value of p, r defectives will be found. Although it is intuitively obvious that about np defectives will occur in the sample, it is not clear, without knowledge of the distribution, how much departure from np is likely to occur just because of the randomness. The binomial density provides this information. Notice that although the conditioning event has not been explicitly stated here it is still relevant. It is the event, H, that a random sample of size n has been taken from a (large) lot in which the proportion defective is p. The distribution is binomial, given H. With other conditioning events it need not be binomial: thus, if the samples were to be taken from different lots with different p's we might obtain another distribution.

Sometimes the samples are taken, not from a lot randomly, but from the production line at regular intervals. The binomial distribution may be relevant to this sampling method, but not necessarily so: for the production process may change with time so that defectives become more or less common, and the successive samples may not be independent, because one defective may alter the probability for the next one taken. In any application

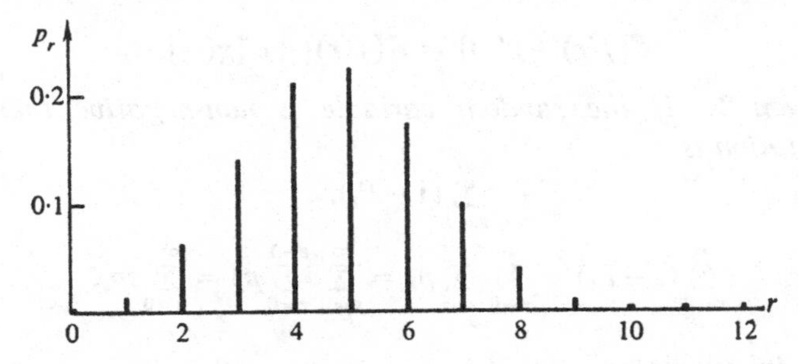

Fig. 2.1.1. The binomial density, $B(12, 0{\cdot}4)$

$$p_r = \binom{12}{r}(0{\cdot}4)^r\,(0{\cdot}6)^{12-r}.$$

of the binomial distribution one should check that the assumptions on which its derivation is based, i.e. of independence and constant p (the hypothesis H above), are reasonably true.

A second, similar example is the inspection of seeds for germination in order to test if they reach the minimum standard laid down by law. A third application is to genetics: for example, in the mating of two heterozygotes the probability of a heterozygote resulting is 1/2, and so the distribution of the number of heterozygotes from n independent matings will have a binomial distribution with index n and parameter 1/2. The number of animals affected in the insulin example (§1.6) will usually have a binomial distribution of index 20 and parameter 1/2 if the insulin is of the correct potency, but otherwise will be different from 1/2. Another application, to the random walk, is given in §2.5.

The name, binomial, is applied to this distribution because (1) is the coefficient of x^r in the expansion of $(px+q)^n$ by the binomial theorem. Fig. 2.1.1 represents the density in the case $p = 0{\cdot}4$, $n = 12$. The abscissa is r and a solid line of height p_r is erected vertically for each r, $0 \leqslant r \leqslant 12$. The probability increases with r up to a maximum (here at $r = 5$) and then diminishes steadily. This is the usual shape, but for p near 0 the maximum may occur at $r = 0$ and p_r steadily diminish with r: similarly, with q near 0 the maximum may occur at $r = n$.

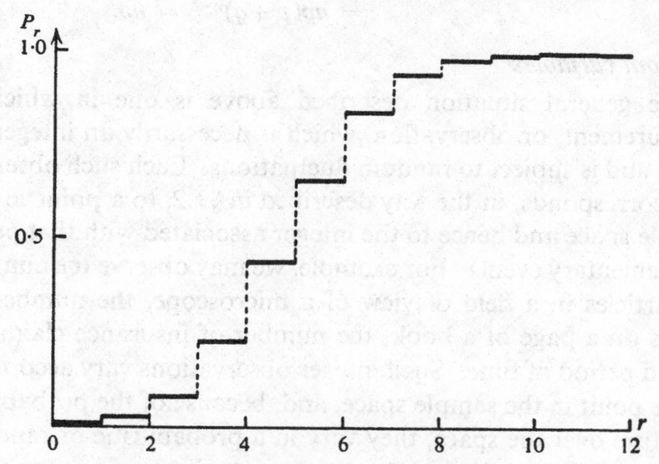

Fig. 2.1.2. The binomial distribution function, B (12, 0·4)

$$P_r = \sum_{s=0}^{r} \binom{12}{s} (0{\cdot}4)^s (0{\cdot}6)^{12-s}.$$

Fig. 2.1.2 represents the distribution function for the same case. r is still the abscissa, and the ordinate at r is P_r, equal to

$$\sum_{s=0}^{r} \binom{n}{s} p^s q^{n-s}.$$

The definition of a distribution function may usefully be extended by defining it as $F(x)$, the probability that the random variable is less than or equal to x, for any real number x, not merely for an integer (cf. §2.2, below). Clearly $F(x) = P_r$ where r is the integral part of x. The figure has been completed by adding the horizontal portions for $F(x)$. Notice that the vertical 'jump' of the function at $x = r$ is p_r, the probability that the random variable equals r.

The mean value of the binomial distribution is easily found by the binomial theorem

$$\sum_{r=0}^{n} rp_r = \sum_{r=0}^{n} r\binom{n}{r} p^r q^{n-r} = \sum_{r=1}^{n} \frac{n!}{(r-1)!(n-r)!} p^r q^{n-r}$$

$$= np \sum_{r=1}^{n} \binom{n-1}{r-1} p^{r-1} q^{(n-1)-(r-1)}$$

$$= np(p+q)^{n-1} = np. \tag{9}$$

Random variables

The general situation described above is one in which a measurement, or observation, which is necessarily an integer, is made and is subject to random fluctuations. Each such observation corresponds, in the way described in §1.2, to a point in the sample space and hence to the integer associated with that point (or elementary event). For example, we may observe the number of particles in a field of view of a microscope, the number of errors on a page of a book, the number of insurance claims in a fixed period of time. Such integer observations vary according to the point in the sample space, and, because of the probability structure over the space, they vary in a probabilistic or random way: hence the name random variable. An integer is an example of a discrete variable. Other discrete variables occur in practice but can nearly always be changed to an integer form. For example, measurements may be made of length to the nearest $\frac{1}{4}$ in., so yielding discrete values: multiplication by 4 will convert them to integers.

We shall find it convenient in the early stages of the discussion to distinguish in the notation between the random variable and the values it takes. Each $a \in \mathrm{A}$ has associated with it an integer $r(a)$. $r(a)$ is the value the random variable takes at a. On the other hand, the random variable is the rule of association which attaches an integer r to each a. This association may be denoted by $r(.)$, reserving $r(a)$ for the value of the random variable at a. The distinction is the same as that between the cosine function which associates with every angle a real value and cos x which is the value of the cosine function for angle x. The cosine function can be written cos(.). The notation $r(.)$ is a little cumbersome and we shall follow Raiffa and Schlaifer (1961) and use $\tilde{r}$. Thus $\tilde{r}$ is the random variable and r is a typical value that it takes. In this notation (5) may be written

$$\mathscr{E}(\tilde{r}) = \Sigma r p_r$$

and (7) becomes

$$\mathscr{E}[f(\tilde{r})+g(\tilde{r})] = \mathscr{E}[f(\tilde{r})]+\mathscr{E}[g(\tilde{r})].$$

The distinction between $\tilde{r}$ and r may assist the understanding in the early stages, but is more conveniently dropped in later work where it does more harm than good. We shall not therefore adhere rigorously to it: only using the tilde when the distinction seems helpful. The situation is analogous to the reference to 'a table of cos x', meaning a table of the cosine function, not merely the value of that function for angle x.

The manner of the random variation can be described in several ways. Two are described above, the density and the distribution function. Others will be met later. Either of these gives the probability that $\tilde{r}$ will assume any given integer value r, the former directly, the latter by minor calculations. The way that the probability is distributed amongst the integers describes the probability distribution of the random variable. Distributions are often considered without any random variable in mind, but merely as a set of numbers satisfying (2). Properties of distributions derived from such considerations can be applied to any random variable having that distribution. Examples of distributions besides the binomial will occur throughout the book.

Expectation

The idea of expectation occurs very early in the study of probability and it is even possible to start with the concept of expectation as the basic idea and develop probability from the properties of expectation, contrary to the method used here. It first arose in games of chance.† Suppose after each play‡ of the

† Like roulette; as distinct from a game of skill, like bridge.
‡ The game ofroulette: a play of roulette is a single throw of the ball.

game a player is given a prize, $\tilde{r}$, supposed integer valued, and $1 \leqslant r \leqslant k$, depending on the outcome of the play. Let A_r be the event of obtaining a prize r and let $p(A_r) = p_r$, so that $\{p_r\}$ is the density of a distribution. If a large number, n, of independent plays result in A_r occurring m_r times, then, as explained in § 1.1, $\text{'lim'}_{n\to\infty} m_r/n = p_r$. The player's total prize after n plays is $\sum_{r=1}^{k} rm_r$ and hence his average prize per play is $\sum_{r=1}^{k} r(m_r/n)$. Now let $n \to \infty$ and with a commonsense interpretation of the limiting operation this will tend to $\sum_{r=1}^{k} rp_r = \mathscr{E}(\tilde{r})$. Hence $\mathscr{E}(\tilde{r})$ is the average prize per play in a long series of plays: it is the amount the player 'expects' to win in each play: it is the amount he should pay the prize-giver before he plays in order that the game be fair. Of course, this explanation is only an interpretation: we shall later (§ 3.6) prove a result, within the mathematical system, which says that the long-run average of a random variable is its expectation.

Expectation has also a meaning in connexion with degrees of belief. The scientist quoted in § 1.6 would expect to gain nothing from a bet at odds of 9 to 1 in favour of H_1, and expect to gain $\frac{1}{2}$ unit from a bet at odds of 4 to 1 with a stake of 1 unit. This is obtained, as (5), by multiplying each prize (one of which is negative) by its probability $(1 \times \frac{9}{10} - 4 \times \frac{1}{10} = \frac{1}{2})$. The generalization to $\mathscr{E}[f(\tilde{r})]$ is needed if, for example, the value of a prize of amount r is equal to $f(r)$. It is not necessary to calculate the distribution of the new prize in order to find its expectation; the latter can be calculated in terms of $\{p_r\}$ directly by (6).

Another interpretation of expectation is of interest. It has already been explained (example 2, § 1.2—and we use the notation of that example) that the elementary events in a discrete sample space may be thought of as particles with masses, m_r, proportional to the probabilities of the events, p_r. If, with each elementary event, is associated an integer r, the particles may be placed at distances r along a line from some convenient origin. This association also defines a random variable, $\tilde{r}$. Let μ be the distance of the centre of gravity of the particles so placed from the origin. Then since the moment about the centre of gravity is zero, $\Sigma(r-\mu)m_r = 0$; but m_r is proportional to p_r, and $\Sigma p_r = 1$, so that $\mu = \Sigma r p_r$. Hence with this placing of the particles the mean is at the centre of gravity. The expectation of $f(\tilde{r})$ can similarly be associated with the centre of gravity when the mass m_r is distant $f(r)$ from the origin.

The reason for insisting on the series (5), if infinite, being absolutely convergent is that the sum of a conditionally convergent series depends on the order in which the terms are taken, and this would conflict with the intuitive concept of expectation which does not involve any notion of ordering. Despite this, if the random variable is non-negative† and (5) diverges we sometimes say that the expectation is $+\infty$; and similarly $-\infty$ with a non-positive random variable. An example is provided in theorem 2: the series (8) diverges iff that for the mean does. We need this result later (in proving theorem 4.4.2).

Geometric distribution

This section is concluded by giving another example of a discrete random variable. Consider again a random sequence of trials with constant probability p of success but, instead of fixing n, the number of trials, and calculating the density of r, take the random variable, $\tilde{s}$, which is the number of trials up to, but not including, the first success. This can only take the value s if the first s trials all result in failures and the $(s+1)$st trial results in a success. Hence. by the multiplication law and the independence of the trials

$$p_s = q^s p \quad (s \geqslant 0). \tag{10}$$

This is the density of a *geometric* distribution with parameter q: we denote it $G(q)$. The reason for the name is that the successive terms have a common ratio q. The mean is

$$\sum_{r=0}^{\infty} r p_r = \sum_{r=0}^{\infty} r q^r p = pq \sum_{r=0}^{\infty} \frac{d}{dq} q^r = pq \frac{d}{dq} \frac{1}{1-q} = q/p. \tag{11}$$

2.2. The continuous case

An extension of the general situation discussed in the previous section is to consider associated with each elementary event a

† That is it never assumes negative values (or does so with probability zero).

real finite number, $\tilde{x}$, not necessarily an integer. Let B_x be the set of a for which $\tilde{x} \leqslant x$, where x is any real number. Write $p(B_x) = F(x)$. Clearly $F(x)$ is non-decreasing, that is

$$F(x) \leqslant F(y) \quad \text{if} \quad x \leqslant y, \tag{1}$$

since the event B_x implies B_y (theorem 1.4.3). It also possesses the properties

$$\lim_{x \to -\infty} F(x) = 0, \quad \lim_{x \to +\infty} F(x) = 1. \tag{2}$$

To prove the first of these statements, we remark that if x is an integer, then, with n also an integer,

$$F(x) = \sum_{n=-\infty}^{x} p(n-1 < \tilde{x} \leqslant n):$$

the series being convergent, we must be able to find a large negative x such that $F(x)$ is arbitrarily small. The second follows similarly on considering $1-F(x)$. A function satisfying (1) and (2) is called a *distribution function*. The number, $\tilde{x}$, associated with each a is called a *random variable* and $F(x)$ is the distribution function of the random variable. The random variable is said to have a *distribution*, and one way of describing it is by the distribution function. These definitions agree with and extend those of §2.1.

The distribution function of §2.1 (in the form with continuous x) had discontinuities at the integer values. Suppose instead that there exists a function $f(x)$ such that

$$F(x) = \int_{-\infty}^{x} f(t)\, dt, \tag{3}$$

then $f(t)$ is called a (*probability*) *density* (of a random variable). Since $F(x)$ satisfies (1) and (2), we have

$$f(x) \geqslant 0, \quad \int_{-\infty}^{\infty} f(x)\, dx = 1. \tag{4}$$

Any function satisfying (4) is called a (probability) density. It follows from the fundamental theorem of the integral calculus that if $f(x)$ is continuous, then

$$f(x) = dF(x)/dx. \tag{5}$$

Let A be the set of a having some property R; then we write $p(A) = p(R)$. For example, if R is the property $\tilde{x} \leqslant x$ we write $p(B_x) = p(\tilde{x} \leqslant x) = F(x)$. This economizes on notation, and $p(\cdot)$ is to be read 'the probability that...'. An immediate consequence of (3) is that

$$F(x_2) - F(x_1) = p(x_1 < \tilde{x} \leqslant x_2) = \int_{x_1}^{x_2} f(t)\,dt, \tag{6}$$

in words, the area under a density between two limits is the probability that a random variable with that density lies between those limits. If the integral

$$\int_{-\infty}^{\infty} xf(x)\,dx \tag{7}$$

is absolutely convergent its value is called the *mean* of the density, μ, or the *expectation* of the random variable $\mathscr{E}(\tilde{x})$. If $g(x)$ is any function and

$$\int_{-\infty}^{\infty} g(x)f(x)\,dx$$

is absolutely convergent, the integral is called the expectation of $g(\tilde{x})$ and written $\mathscr{E}[g(\tilde{x})]$. Theorem 2.1.1 extends to this case.

The analogue of theorem 2.1.2 is

Theorem 1. *If the random variable, with density $f(x)$, is non-negative then its expectation is*

$$\int_0^{\infty} \{1 - F(x)\}\,dx. \tag{8}$$

For

$$\begin{aligned} \int_0^{\infty} \{1 - F(x)\}\,dx &= \int_0^{\infty} \left\{1 - \int_0^x f(t)\,dt\right\} dx \\ &= \int_0^{\infty} dx \int_x^{\infty} dt\, f(t) = \int_0^{\infty} dt \int_0^t dx\, f(t) \\ &= \int_0^{\infty} tf(t)\,dt. \end{aligned}$$

Continuous random variables

All observations in the real world are discrete because they are made to the limits of accuracy of the measuring instrument, for example to the nearest tenth of a second. But there are considerable mathematical advantages in introducing the idea of continuity and the field of probability is no exception. The real difference between the discrete and continuous cases is that in the former no finer measurement is imagined (there cannot be $3\frac{1}{2}$ successes, for example), whereas in the latter refinements are conceptually possible by using a more accurate apparatus.

The definition given in this section of a distribution function, equations (1) and (2), is general and includes the discrete case: namely, it is the probability that the random variable is less than or equal to a given number. Any random variable has a distribution function and it is a remarkable fact, that we shall not prove, that given the distribution function the probability of *any* event concerning solely the random variable can be calculated: we have seen how this can be done for the probability that it lies between two limits (equation (6)). Thus the distribution function completely defines the distribution of the random variable. In the discrete case we saw that the distribution function was constant in intervals with integer end-points and was discontinuous, or had 'jumps' at integer values equal to the probabilities of the random variable being equal to those integer values. On the other hand, if $F(x)$ can be written as an integral (equation (3)), it is necessarily continuous everywhere since an indefinite integral is a continuous function. The converse is not true, namely a continuous distribution function cannot necessarily be written in the form of (3), but the possibility of a continuous $F(x)$ not satisfying (3) can be ignored in most applications. An example will be given in §3.2. The case where (3) obtains will be referred to as the *continuous case*, and we shall refer to a random variable with a density in the sense of this section as a *continuous* random variable.

Probability density

To help understand the meaning of the density and equation (5) in the continuous case, an analogy from mechanics is again useful. Consider a rod such that the total mass of the rod to the left of x is $F(x)$. (If $\tilde{x}$ can take arbitrarily large values the rod will be infinite in extent: consideration of the case where necessarily $x_1 \leqslant \tilde{x} \leqslant x_2$ so that we have a finite rod of length

$x_2 - x_1$ may be easier.) The case of a uniform rod was given in example 1 in §1.2. In the general case what is the mass density of the rod at x? Density is defined by taking a small interval about x, dividing the mass of the rod in the interval by the width of the interval and allowing the latter to approach zero. If the interval $(x, x+h)$ is taken the ratio is $\{F(x+h)-F(x)\}/h$ which tends to $f(x)$ (equation (5)). Hence, in the analogy, $f(x)$ is the mass density of the rod, and the term probability density is natural for the general case. Consequently the probability that a continuous random variable lies in a small interval of width h about x is approximately $f(x)h$. Sometimes $f(x)$ is defined by (5), but this, besides being wrong in general, misses the main function of $f(x)$, namely its use in (6): the density is a function which, when integrated over a region (and in particular an interval), gives the probability of the random variable belonging to that region. Notice that *density* has slightly different meanings in the discrete and continuous cases. The justification for a common name is that the two functions behave in exactly the same way in most cases, as we shall see repeatedly below.

Notice that we have supposed a random variable to be finite. Thus, if $\tilde{r}$ is a random variable with non-zero probability of being itself zero, then $1/\tilde{r}$ is not a random variable. Our random variables are often said to be *honest*.

Although most distribution functions will either be discrete (§2.1) or continuous, we shall sometimes meet the *mixed* case. An example (§4.2) is the time a customer has to wait to be served by a shop-assistant. If the shop-assistant is free when the customer enters the shop the time will (with an attentive assistant) be zero. However, if there are other customers present the new arrival will have to wait until they have been attended to. The waiting-time is measured continuously (from 0 to $+\infty$) but, if p is the probability that the assistant is free, p will typically (for the assistant's sake) be non-zero. Hence we can expect $F(x)$ to be zero for $x < 0$, be equal to p at $x = 0$ and be continuous for $x > 0$. It cannot therefore be written in the form (3) but it can be written for $x \geqslant 0$ as

$$F(x) = p + \int_0^x f(t)\,dt,$$

and with obvious modifications can be handled like a continuous distribution function. Notice generally that if $F(x)$ is discontinuous at $x = x_0$ the 'jump' at the discontinuity is the probability that the random variable equals x_0: conversely, if $F(x)$ is continuous the probability that the random variable equals x_0 is zero. This last result explains why we could not proceed, as in §2.1, by defining an event A_x containing all elementary events for which the random variable equals x and using $p(A_x)$: it would, in the continuous case, be zero.

Expectation

The definition by (7) of the mean or expectation is a natural generalization of the corresponding definition (equation 2.1.5) in the discrete case. Suppose the range of $\tilde{x}$ divided into small intervals of equal widths δx. The probability of $\tilde{x}$ lying in one of these intervals is approximately $f(x_r)\,\delta x$ (by what we have just discussed), where x_r is some point of the interval. Hence $f(x_r)\,\delta x$ may be compared with p_r and the mean value is approximately $\sum_r x_r f(x_r)\,\delta x$, which is an approximating sum to the integral $\int_{-\infty}^{\infty} xf(x)\,dx$. The justification for the integral for $\mathscr{E}[g(\tilde{x})]$ is similar. As in the discrete case μ is the position of the centre of gravity of the rod just described.

The idea of replacing a summation (Σrp_r) by an integration $(\int xf(x)\,dx)$ in passing from the discrete case to the continuous one is of frequent use. The proof of theorem 1 is an example of such a change from that of theorem 2.1.2. We shall often give a proof of a theorem that covers only one of the cases and then remark that the other case follows on changing from summation to integration or vice versa. There is a mathematical technique (Stieltje's integration) which unites the two cases, but it is beyond the mathematical level of this book. Notice that the density, $\{p_r\}$ or $f(x)$, is used similarly in the two definitions of the expectation: this is the reason for using the same name to describe $\{p_r\}$ and $f(x)$. In order to give interesting examples of continuous distributions we pass, in the next section, to the consideration of an important probability problem.

2.3. The Poisson process

Consider a sample space in which each elementary event consists of an infinite sequence of real numbers $\{t_i\}$ with $0 \leqslant t_1 < t_2 < \ldots < t_n < t_{n+1} < \ldots$. The elementary event corresponds to the observation of a process, beginning at time $t = 0$, during which any number of *incidents* can occur; the rth incident occurring at time t_r. Let t and h be such that $0 \leqslant t < t+h$. Let A be any event which refers only to incidents in the interval $(0, t)$ and, for any non-negative integer k, let B be the event of k incidents in the interval $(t, t+h)$.† If the events A and B, so defined, are always independent the probability system is said to be a *purely random process*. If $p(B \mid A)$, which, in virtue of the independence, may be written $p(B)$, depends only on h and k, and not on t, the process is said to be *stationary* and the system is said to be a *purely random stationary process*, or a *Poisson process*.

Theorem 1. *In a Poisson process,* $p_0(t)$, *the probability of no incidents in a fixed interval of length* t, *is* $e^{-\lambda t}$, *where* λ *is non-negative.*

Consider an interval of length $t+h$. Because the process is stationary it does not matter where the interval begins. No incidents occur in it iff none occur in either of the non-overlapping subintervals of lengths t and h. These last two events are independent, by the definition of a Poisson process. Hence by the multiplication law (equation 1.3.2),

$$p_0(t+h) = p_0(t)\, p_0(h). \tag{1}$$

Taking logarithms of each side shows that $\ln p_0(t)$‡ is an additive function of its argument, and so is $-\lambda t$, for some λ. Finally, λ must be non-negative since only probability solutions, that is solutions in $(0, 1)$, are meaningful.

Theorem 2. *In a Poisson process the density,* $f_1(x)$, *of the time,* $\tilde{t}_1$, *to the first incident is* $\lambda e^{-\lambda x}$, *for* $x \geqslant 0$. *For* $x < 0$ *it is obviously zero.*

† The instant t may belong to either, but not both, of the intervals.

‡ Throughout this book we use natural logarithms, to the base e, and denote them by ln.

Let x be any positive number. Then $\tilde{t}_1 > x$ iff no incidents occur in $(0, x)$. But by theorem 1 this probability is $e^{-\lambda x}$. Hence

$$p(\tilde{t}_1 \leqslant x) = 1 - e^{-\lambda x}, \tag{2}$$

and the left-hand side is, by definition, the distribution function of $\tilde{t}_1$, which clearly has a density, obtained by differentiation; namely

$$f_1(x) = \lambda e^{-\lambda x}. \tag{3}$$

Corollary. *In a Poisson process the density of the time between any two successive incidents is also given by (3).*

Consider, for example, $\tilde{t}_2 - \tilde{t}_1 = \tilde{s}$, say, and consider its distribution when the conditioning event is that $\tilde{t}_1$ is some fixed value, u. That is, consider the Poisson process when the first incident happens at u. The basic assumptions of independence of events concerning the intervals $(0, u)$ and $(u, u+x)$ and the stationarity mean that the same arguments used in the proof of the theorem for the interval $(0, x)$ apply to the interval $(u, u+x)$. Hence the density is still given by (3). Since this does not involve u, $\tilde{s}$ must be independent† of $\tilde{t}_1$, and hence (3) is the density for general u.

Equation (3) is our first example of a density: it is referred to as the density of the *exponential distribution* with *parameter* λ. It is a special case of the Γ-distribution below. The distribution will be denoted by $E(\lambda)$.

Theorem 3. *Let t be any fixed positive number. In a Poisson process, $p_n(t)$, the probability of n incidents in an interval of length t, is $e^{-\lambda t}(\lambda t)^n/n!$.*

Consider‡ any interval of length t, say $(0, t)$ and divide it up into a large number, N, of intervals each of length $\delta s = t/N$, let m be an integer, $1 \leqslant m \leqslant N$, and $s = mt/N$. Then the event of n (>0) incidents in $(0, t)$ can take place in one of the following N exclusive and exhaustive ways: the first incident occurs in the interval $[(m-1)t/N, mt/N]$ and there are $(n-1)$ incidents in the interval $[mt/N, t]$, which two events are independent. Now if N

† The independence of two random variables, here $\tilde{s}$ and $\tilde{t}_1$, has not been defined. This will be done in §3.2. Here we mean by 'independent of $\tilde{t}_1$', 'the distribution does not depend on the value of $\tilde{t}_1$', namely u.

‡ The proof contains a minor gap. This will be discussed below.

is large so that $t/N = \delta s$ is small, the first probability is approximately, by (3) and the general property of a density, $\lambda e^{-\lambda s}\delta s$; that is, the density at some point (here $s = mt/N$) of the interval times the width of the interval. The second probability is $p_{n-1}(t-s)$. Hence, by the addition law, $p_n(t)$ is approximately

$$\sum_{m=1}^{N} \lambda e^{-\lambda s} p_{n-1}(t-s)\,\delta s \quad (s = mt/N). \tag{4}$$

But this is an approximating sum to an integral so that, on allowing N to tend to infinity ($\delta s \to 0$), we have

$$p_n(t) = \int_0^t \lambda e^{-\lambda s} p_{n-1}(t-s)\,ds. \tag{5}$$

The result now follows by induction. If

$$p_{n-1}(t-s) = e^{-\lambda(t-s)}\lambda^{n-1}(t-s)^{n-1}/(n-1)!$$

then, by (5),

$$p_n(t) = e^{-\lambda t}\lambda^n \int_0^t (t-s)^{n-1}\,ds/(n-1)!$$

$$= e^{-\lambda t}(\lambda t)^n/n!. \tag{6}$$

Hence if the result to be proved is true for $(n-1)$ it is true for n. But it is true for $n = 0$ (theorem 1). Hence it is true generally.

$\{p_n(t)\}$, for fixed t, is a probability density (of the random variable $\tilde{n}$). It is called the density of the *Poisson distribution.* It depends only on the product $\lambda t = \mu$, say: μ is called the *parameter* of the distribution and we write $P(\mu)$ for a Poisson distribution with parameter μ.

Theorem 4. *The expected number of incidents in any time interval of length t is λt.*

The probability distribution of the number is $P(\lambda t)$ so we have only to find the mean of the Poisson distribution. This is, where $\mu = \lambda t$,

$$\sum_{n=0}^{\infty} np_n(t) = e^{-\mu}\mu \sum_{n=1}^{\infty} \mu^{n-1}/(n-1)! = \mu,$$

since the series is the exponential series with sum e^{μ}.

This theorem enables a physical interpretation to be given of λ: it is the expected number of incidents per unit time. It is the *rate* of occurrence of the incidents. (See also the discussion below).

Theorem 5. *In a Poisson process, if n is a fixed number, the time, $\tilde{t}_n$, from the start to the n-th incident has a density $f_n(x)$ given by*

$$f_n(x) = e^{-\lambda x}\lambda^n x^{n-1}/(n-1)! \tag{7}$$

for $x \geqslant 0$ and zero for $x < 0$.

The theorem is a generalization of theorem 2, and the proof is similar to the proof of that theorem. Let x be any positive number. Then $\tilde{t}_n > x$ iff $(n-1)$, or fewer, incidents occur in $(0, x)$. But by theorem 3 and the addition law this has probability

$$e^{-\lambda x}\{1 + (\lambda x) + (\lambda x)^2/2! + \dots + (\lambda x)^{n-1}/(n-1)!\}. \tag{8}$$

Hence $p(\tilde{t}_n \leqslant x)$, the distribution function, is 1 minus this expression. Differentiation of that function gives, because most of the terms cancel in pairs, the result (7).

Corollary. *In a Poisson process the density of the time between two incidents which are n apart is also given by* (7).

The proof is similar to the proof of the corollary to theorem 2.

The density (7) is said to be the density of a *gamma distribution*. λ is called the *parameter* and n the *index*. We write $\Gamma(n, \lambda)$ for a gamma (Γ) distribution of index n and parameter λ. The exponential distribution (above) is $\Gamma(1, \lambda)$.

Theorem 6. *The expectation of the time up to the n-th incident, $\tilde{t}_n$, is n/λ.*

The probability distribution of the time is $\Gamma(n, \lambda)$ so we have only to find the mean of the Γ-distribution. This is (equation 2.2.7)

$$\int_0^\infty x f_n(x)\,dx = \int_0^\infty e^{-\lambda x}(\lambda x)^n\,dx/(n-1)!.$$

Substitute $\lambda x = t$, so that $\lambda\,dx = dt$. Then the expression is

$$\lambda^{-1}\int_0^\infty e^{-t}t^n\,dt/(n-1)!.$$

The integral is well-known to be $n!$ and the result follows.

This theorem gives a second interpretation of λ. With $n = 1$ the expected (or average) time between successive events is λ^{-1}.

Examples

There are many practical situations where observations are made on incidents: that is, happenings that occupy only an incident of time and are not spread over an interval. A famous example is a Geiger counter, which counts the incidents of the arrival of atomic particles at the counter. Other incidents are the breakdowns of a machine, accidents to a worker, demands for connexion from telephone subscribers, aircraft arriving at an airport and requesting permission to land and, generally, arrivals of 'customers' requiring 'service' at a 'counter'. Such incidents form a process and if the process is subject to probability laws it is called a random or *stochastic* process. The most important feature of any stochastic process is the way in which the past behaviour of the process influences the future behaviour. The purely random process is one in which this influence is nil: that is, the future is independent of the past. If, in the formal definition above, t is thought of as the present, the definition says that whatever happened in the past (that is, whatever A is) does not affect the probability of an incident in the future. It might be thought that such processes are too random, or too chaotic, either to occur in practice or, if they do, to be usefully studied. But this is not so. The Geiger counter provides an excellent example: the individual particles behave quite independently of each other in many situations, the only slight difficulty resides in the fact that the counter has a 'dead time' immediately after a particle has arrived during which it cannot record the arrival of another. But even then the basic arrival process is purely random and from its properties one can deduce the properties of the recorded process. Another excellent example is the telephone one: subscribers act without knowledge of each other's behaviour and there is an extensive and useful theory of the capacity of telephone exchanges based on the theory of the Poisson process. The other examples may sometimes be purely random, but not always. Thus the aircraft may tend to arrive in groups because of weather conditions, or they may arrive rather regularly because they have been scheduled to do so. A newly repaired machine may be more or less reliable than one that has not broken down for a while, and therefore the knowledge of its age since last repair (that is, an event of the type of A) may influence, and not be independent of, a breakdown in the future (this will be discussed in §4.4). Nevertheless, the purely random process provides the foundations for many stochastic processes (for example the queueing process, §§4.2, 4.3). It has been shown that when incidents are of several different types, each type occurring with some degree of regularity, the over-all picture of incidents may be random.

The stationarity restriction means that the origin of time is irrelevant. For example, the telephone process is clearly not stationary, the probability of no calls between 02.00 and 02.05 h is clearly greater than between 14.00 and 14.05 h: nevertheless, it may be stationary over short periods of time, say between 14.00 and 17.00 h. In all these examples the variable, t, has been referred to as time, but it may represent a spatial variable. For example, consider a crop growing in a row. The $\{t_i\}$ may be the distances from the end of the row of points of infection, and the infection is then said to be purely randomly distributed along the row. The notion can also be extended to processes in the plane or in space: the basic requirement is that the probability of an incident in a region is independent of any event concerning incidents outside the region. Flaws in a sheet of metal, dust particles on a plate, raisins in a cake, are three examples where such a probability structure might apply.

Exponential distribution

Theorem 2 provides our first example of a density in the continuous case. The reader should sketch the distribution function (equation (2)) and the density (equation (3)) and consider how they depend on λ. The former is continuous so there are no single values of non-zero probability. The density is discontinuous at the origin, changing from zero to λ, and then diminishing steadily with time. The occurrence of a maximum of the density at $t = 0$ is interesting because it means that the small intervals between successive events are more probable than the longer ones. In a real process (such as telephony) which is purely random, there is more clustering together of the incidents than most people expect. If they are asked to put points at random on a line, say (that is, a spatial case), they tend to distribute them too uniformly. Perhaps this clustering is what lies behind the proverb 'It never rains, but it pours'. An interesting example is the position of fall of flying-bombs on London during the last war. Some places were described as getting an unfair share of the bombs: in fact the distribution was just what one would expect from a Poisson process. An interesting consequence of the purely random property is that you can expect to wait as long for an incident whether one has just occurred or not. This follows from theorem 2 and its corollary.

There is a close connexion between the exponential distribution and the geometric distribution (§2.1). The geometric distribution was obtained by considering a random sequence of trials with constant probability of success and taking the number of trials up to, but not including, the first success. If we think of a success as an incident which can only happen at discrete, equally spaced time points (the trials) then, in virtue of the independence of the trials and the constancy of success at each one, we have a situation analogous to the Poisson process wherein the events in different time-intervals are independent and the chance of an event constant. Consequently the number of trials before the first success is analogous to the time up to the first incident. The densities both have the property that an increase in the value of the random variable from s to $s+m$ causes the density to be divided by a quantity which depends only on m and not on s.

Poisson distribution

Theorem 3, the proof of the Poisson distribution, is an important result. It applies to the number of incidents in a fixed period of time: the number of telephone calls, the number of atomic particles, etc. Fig. 2.3.1 shows the densities (using the same method as with the binomial in §2.1) for μ, the mean, equal to 1 and 3. Fig. 2.3.2 similarly shows the distribution functions. For all but small μ the density increases to a maximum with n and then diminishes. The increase is more rapid than the decrease and the right-hand '*tail*' of the density is longer than the left-hand one. For small μ the maximum is at $n = 0$. It is worth remembering, though the proof is beyond the level of this book, that the Poisson distribution characterizes the Poisson process. That is, not only does the process yield the distribution but the distribution, with mean proportional to the length of the interval, can only arise from the process. A common way of testing whether a process is Poisson is to test whether the number of incidents has a Poisson distribution (§7.5) and then use the characterization.

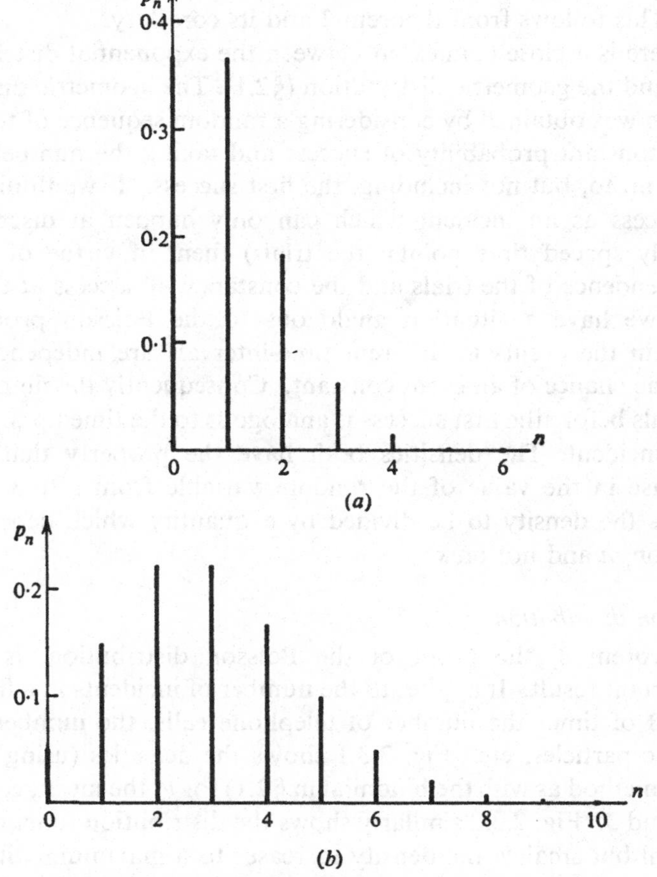

Fig. 2.3.1. (*a*) The Poisson density, $P(1)$ $p_n = e^{-1}1^n/n!$.
(*b*) The Poisson density, $P(3)$ $p_n = e^{-3}3^n/n!$.

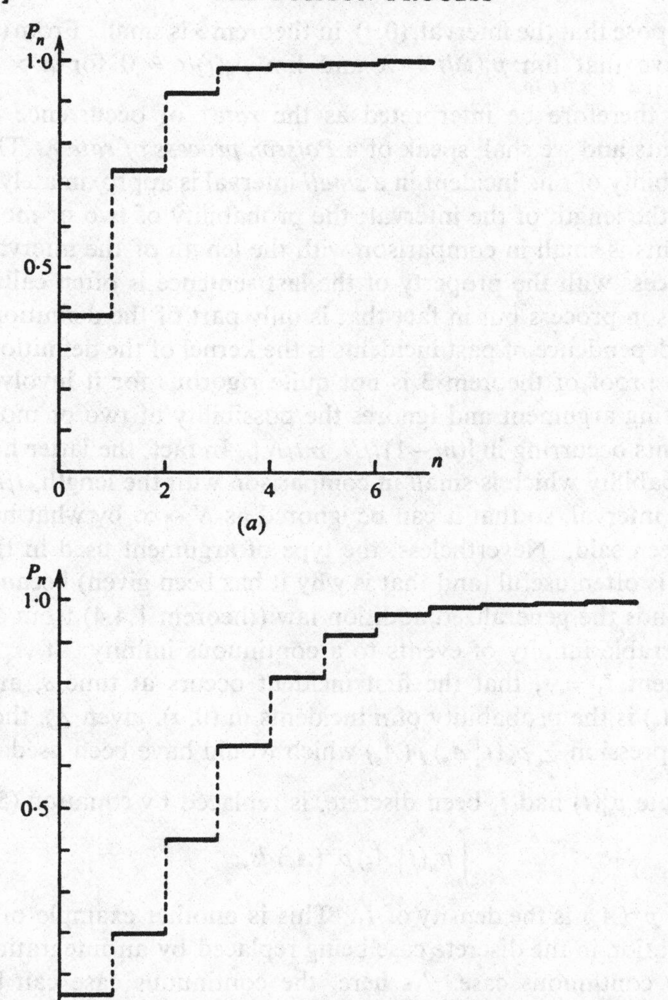

Fig. 2.3.2. (*a*) The Poisson distribution function, $P(1)$ $P_n = \sum_{s=0}^{n} e^{-1}1^s/s!$.

(*b*) The Poisson distribution function, $P(3)$ $P_n = \sum_{s=0}^{n} e^{-3}3^s/s!$.

Suppose that the interval, $(0, t)$, in theorem 3 is small. From (6) we have that $\lim_{t \to 0} p_1(t)/t = \lambda$ and $\lim_{t \to 0} p_n(t)/t = 0$ for $n > 1$. λ can therefore be interpreted as the *rate*† of occurrence of incidents and we shall speak of a *Poisson process of rate* λ. The probability of one incident in a *small* interval is approximately λ times the length of the interval; the probability of two or more incidents is small in comparison with the length of the interval. A process with the property of the last sentence is often called a Poisson process but in fact that is only part of the definition: the independence of past incidents is the kernel of the definition.

The proof of theorem 3 is not quite rigorous for it involves a limiting argument and ignores the possibility of two or more incidents occurring in $[(m-1)t/N, mt/N]$. In fact, the latter has a probability which is small in comparison with the length, t/N, of the interval, so that it can be ignored as $N \to \infty$ by what has just been said. Nevertheless, the type of argument used in the proof is often useful (and that is why it has been given) because it extends the generalized addition law (theorem 1.4.4) from an enumerable infinity of events to a continuous infinity. If A_s is the event $\tilde{t}_1 = s$, that the first incident occurs at time s, and $p_n(t \mid A_s)$ is the probability of n incidents in $(0, t)$, given A_s, then the expression $\sum_s p_n(t \mid A_s)\, p(A_s)$ which would have been used to calculate $p_n(t)$ had $\tilde{t}_1$ been discrete, is replaced by equation (5)

$$\int p_n(t \mid A_s)\, p^*(A_s)\, ds,$$

where $p^*(A_s)$ is the density of $\tilde{t}_1$. This is another example of a summation in the discrete case being replaced by an integration in the continuous case. As here, the continuous case can be reached by considering the discrete case first, with the probabilities being products of densities and lengths of small intervals of the random variable, and then passing to a limit as these lengths tend to zero: the discrete sums are approximating sums to the integral limit. This will be discussed further in §3.2.

A rigorous proof of the Poisson distribution goes along the

† Compare the concept of velocity (or rate of motion). A particle has velocity v if the distance covered in a small interval is approximately v times the length of the interval.

following lines. Divide the interval $(0, t)$ into two intervals $(0, u)$ and (u, t). Then using the addition law and the basic independence property of the process we have

$$p_1(t) = p_0(u)\,p_1(t-u)+p_1(u)\,p_0(t-u)$$

(cf. equation (1)). Since $p_0(t)$ is known this is an equation for $p_1(t)$. Generally $p_n(t)$ can be written as an expression involving $\{p_r(t)\}$ for $r < n$. Successive solutions of these equations give the required result (cf. also example 2.6.3). Notice that we have tacitly made the assumption that the incidents are ordered. That is, at any point of time there is a next incident, the first to occur after that point. Thus the possibility of two incidents occurring at the same time is not admitted, nor is the more recondite possibility of them occurring at times n^{-1} $(n = 1, 2, \ldots)$ so that there is no next incident at $t = 0$.

Alternative derivation of the Poisson distribution

The Poisson distribution is named after the French mathematician, Poisson, but he derived it in a different way, namely as a limit of the binomial distribution. We now give the modern version of his derivation, although it amounts to a third proof of theorem 3, because it is still of some interest. Let $\{p_r\}$ be the density of a binomial distribution $B(n, p)$. The mean is np (equation 2.1.9) so that if we allow n to approach infinity and p to approach zero with $np = \mu$, constant, we might expect the distribution to tend to a limit since its mean will remain finite. With $p = \mu/n$, (2.1.1) becomes, with some rearranging,

$$p_r = \frac{\mu^r}{r!}\frac{n(n-1)\ldots(n-r+1)}{n^r}\left(1-\frac{\mu}{n}\right)^n\left(1-\frac{\mu}{n}\right)^{-r}.$$

Now allow n to tend to infinity. The first term does not involve n, so remains $\mu^r/r!$; the second tends to 1 since it is the product of r terms each of which tends to 1; the third tends to $e^{-\mu}$ by a standard result; the fourth term tends to $1^{-r} = 1$. Hence the limit of p_r is $e^{-\mu}\mu^r/r!$, the density of the Poisson distribution. The practical significance of this is that for binomial situations with large n and small p the more easily calculated $P(\mu)$ may replace $B(n, p)$ if $\mu = np$. A classic, but not very good, example is deaths from the kicks of horses in the Prussian cavalry. The number of exposures, n, to being kicked was large, but the chance of death when exposed, p, small. But it is possibly more realistic to think of it as a Poisson process with incidents (death from the kick of a horse) occurring independently of past incidents. The formal connexion between the proof just given and the process is obtained by dividing the interval $(0, t)$ into n small intervals. The probability of an incident in a small interval of length t/n is $\lambda t/n = p$ and, because of the independence and stationarity, the n intervals are like a random sequence of n trials with constant probability p of success. Hence for finite n the distribution is binomial $B(n, p)$ and $p = \lambda t/n$. Allow $n \to \infty$, λ and t fixed, apply the above result and we have a proof (not quite rigorous again) of theorem 3.

The gamma distribution

The gamma distribution of theorem 5 will occur again later (example 3.5.1; §5.3). Meanwhile we remark that the density (for $n > 1$) has the value zero at $x = 0$, increases to a maximum at $x = (n-1)/\lambda$ and then decreases: the rate of decrease being slower than the increase. Notice that this distribution is of the random variable, 'the time taken for a fixed number of incidents'; in contrast to the Poisson distribution, which is of the random variable, 'the number of incidents in a fixed time'. In other contexts the index of the Γ-distribution need not be an integer. The function $f_n(x)$ in (7) represents a density; that is, satisfies equation 2.2.4, provided $n > 0$ and $n!$ is defined by $\int_0^\infty e^{-t}t^n\,dt$. This integral is usually called the gamma function and is denoted by $\Gamma(n+1)$; hence the name of the distribution. The Γ-notation for the integral will not be used in this book because of the confusion over the $n+1$ in the argument when n is the power of t in the integrand. The integral will be called the *factorial function*. $(-\frac{1}{2})!$ will be evaluated in §2.5.

2.4. Features of distributions

The *mean* of a distribution has already been defined (§§2.1, 2.2) as the expectation of a random variable with that distribution. The elementary results

$$\mathscr{E}(c\tilde{x}) = c\mathscr{E}(\tilde{x}), \quad \mathscr{E}(\tilde{x}+c) = \mathscr{E}(\tilde{x})+c, \tag{1}$$

where c is a constant and $\tilde{x}$ a random variable, discrete or continuous, follow immediately from the definitions. The *variance* of a distribution is defined†, provided that the expressions converge, as

$$\int (x-\mu)^2 f(x)\, dx \quad \text{or} \quad \sum_r (r-\mu)^2 p_r \tag{2}$$

in the continuous and discrete cases respectively, and is usually denoted by σ^2. Here μ is the mean. If $\tilde{x}$ is a random variable with a distribution of variance σ^2 we speak of $\tilde{x}$ as having variance σ^2 and write it $\mathscr{D}^2(\tilde{x})$. From equations 2.1.6 or 2.2.7 and (2) above it follows that

$$\sigma^2 = \mathscr{D}^2(\tilde{x}) = \mathscr{E}[\{\tilde{x}-\mathscr{E}(\tilde{x})\}^2], \tag{3}$$

expressing the variance in terms of expectations. From (1) and (3) it follows that

$$\mathscr{D}^2(c\tilde{x}) = c^2\mathscr{D}^2(\tilde{x}), \quad \mathscr{D}^2(\tilde{x}+c) = \mathscr{D}^2(\tilde{x}). \tag{4}$$

The positive square root, σ or $\mathscr{D}(\tilde{x})$, is called the *standard deviation* of the distribution or of the random variable.

Theorem 1.‡

$$\mathscr{D}^2(\tilde{x}) = \mathscr{E}(\tilde{x}^2) - \mathscr{E}^2(\tilde{x}). \tag{5}$$

For $\mathscr{D}^2(\tilde{x}) = \mathscr{E}[\tilde{x}^2 - 2\tilde{x}\mathscr{E}(\tilde{x}) + \mathscr{E}^2(\tilde{x})]$, by (3),

$= \mathscr{E}(\tilde{x}^2) - 2\mathscr{E}^2(\tilde{x}) + \mathscr{E}^2(\tilde{x})$, by (2.1.7),

as required.

A slightly different version of this theorem is useful for some discrete distributions. The proof is immediate from theorem 1.

Theorem 2.

$$\mathscr{D}^2(\tilde{x}) = \mathscr{E}[\tilde{x}(\tilde{x}-1)] + \mathscr{E}(\tilde{x}) - \mathscr{E}^2(\tilde{x}). \tag{6}$$

Theorem 3 (*Chebychev's inequality*). *If $\tilde{x}$ is a random variable for which $\mathscr{E}(\tilde{x}^2)$ exists and c is a positive constant, then*

$$p(|\tilde{x}| \geqslant c) \leqslant \mathscr{E}(\tilde{x}^2)/c^2. \tag{7}$$

† Integrals and sums whose ranges are from $-\infty$ to $+\infty$ will be written without the limits.

‡ $\mathscr{E}^2(\tilde{x})$ means $\{\mathscr{E}(\tilde{x})\}^2$.

The proof is given for the continuous case, with density $f(x)$.

$$\begin{aligned}\mathscr{E}(\tilde{x}^2) &= \int x^2 f(x)\,dx, \quad \text{by definition,}\\ &\geqslant \int_{|x|\geqslant c} x^2 f(x)\,dx, \quad \text{since the integrand is positive,}\\ &\geqslant c^2 \int_{|x|\geqslant c} f(x)\,dx, \quad \text{since } |x| \geqslant c \text{ in the range of integration,}\\ &= c^2 p(|\tilde{x}| \geqslant c),\end{aligned}$$

from the basic property of the density (equation 2.2.6).

Corollary. *If μ and σ are respectively the expectation and standard deviation of $\tilde{x}$,*

$$p(|\tilde{x}-\mu| \geqslant c\sigma) \leqslant c^{-2}. \tag{8}$$

This follows from (7) on replacing $\tilde{x}$ by $\tilde{x}-\mu$ and c by $c\sigma$.

Measures of location

The probability structure of a random variable is described by its distribution function, or something equivalent to it such as a density. These are relatively complex descriptions, being functions, but they completely describe the probability structure. In this section we discuss certain numbers which describe *features* of the probability structure but do not characterize it. That is, a whole function is replaced by a single number; the loss of information being compensated for by a gain in simplicity: several distributions having the same number associated with them. The mean, or expectation, is the most important feature and its interpretation has already been discussed. In virtue of the property in (1) of increasing by c if the random variable is increased by c, it is called a *measure of location* of the distribution: it describes the centre or location of the distribution (see also §5.3). Other measures of location sometimes used are the median and the mode. A *median* of a distribution is a value x such that, if $\tilde{x}$ is a random variable with that distribution, $p(\tilde{x} \leqslant x) \geqslant \frac{1}{2}$ and $p(\tilde{x} \geqslant x) \geqslant \frac{1}{2}$. If $F(x)$ is continuous it is a root of the equation $F(x) = \frac{1}{2}$, and the random variable is equally likely to be above or below the median. If

$F(x)$ is never constant in an interval the median is unique. A *mode* of a distribution is a value for which the density has a local maximum. Most distributions have a single local maximum which is the greatest value, such distributions are *unimodal*: all distributions so far encountered are unimodal. One occasionally meets distributions with two modes; they are called *bimodal*.

Measures of spread

In contrast the variance (or standard deviation) measures the scatter of the distribution about the central value, in this case the mean. It is the expectation of the square of the departure of the random variable from the mean. If the random variable is constant there is no scatter and the variance is zero: the more the scatter, the larger the variance. Notice, from (4), that the variance has the dimension of the square of the random variable whereas the standard deviation has the same dimension. The variance is the easier to handle mathematically but the standard deviation is the easier to interpret. In calculations (numerical or algebraic) we usually work with the variance and only at the end convert it into a standard deviation (cf. §§3.3, 5.1). In virtue of its properties in (4) the standard deviation is called a *measure of spread*. Other measures of spread sometimes used are the mean deviation, the range and the inter-quartile range. The *mean deviation* is defined as

$$\int |x-\mu| f(x)\,dx \quad \text{or} \quad \sum_r |r-\mu| p_r,$$

that is $\mathscr{E}[|\tilde{x}-\mu|]$, but the moduli signs make it awkward to handle. The *range* is only applied to a variable which always lies between two finite limits and is the least difference between the limits. For a continuous, strictly increasing distribution function the *quartiles* are the roots x_1 and x_2 of the equations $F(x_1) = \frac{1}{4}$, $F(x_2) = \frac{3}{4}$ and the *inter-quartile range* is (x_2-x_1). A random variable is equally likely to lie between or outside the quartiles. Modifications have to be made, like those for the median, with general distribution functions.

We illustrate the use of theorem 1 by computing the variance of the Γ-distribution. We have already seen (theorem 2.3.6) that its mean is n/λ, so $\mathscr{E}(\tilde{x}) = n/\lambda$, if $\tilde{x}$ has a Γ-distribution. Then

$$\mathscr{E}(\tilde{x}^2) = \int_0^\infty x^2 f_n(x)\,dx = \int_0^\infty e^{-\lambda x}\lambda^n x^{n+1}/(n-1)!,$$

and the substitution $\lambda x = t$ gives $\mathscr{E}(\tilde{x}^2) = n(n+1)/\lambda^2$. Hence, by theorem 1,

$$\mathscr{D}^2(\tilde{x}) = n(n+1)/\lambda^2 - (n/\lambda)^2 = n/\lambda^2. \tag{9}$$

In terms of the mechanical analogy that has been used before (§2.1) in discussing the mean, which was then the centre of gravity of a mass distribution, the variance is the moment of inertia about the centre of gravity. Theorem 1 is the familiar result which enables this moment to be calculated from the moment about the origin, $\mathscr{E}(\tilde{x}^2)$, by subtracting the moment about the origin of the total mass situated at the centre of gravity, $\mathscr{E}^2(\tilde{x})$.

In the discrete case theorem 2 is more useful and we use the binomial for illustration. $\mathscr{E}(\tilde{r}) = np$, if $\tilde{r}$ is $B(n, p)$ (equation 2.1.9). Then (compare the derivation of 2.1.9)

$$\begin{aligned}\mathscr{E}[\tilde{r}(\tilde{r}-1)] &= \sum_{r=0}^{n} r(r-1)\binom{n}{r}p^r q^{n-r}\\ &= n(n-1)p^2\sum_{r=2}^{n}\binom{n-2}{r-2}p^{r-2}q^{(n-2)-(r-2)}\\ &= n(n-1)p^2(p+q)^{n-2} = n(n-1)p^2.\end{aligned}$$

Hence, by theorem 2,

$$\mathscr{D}^2(\tilde{r}) = n(n-1)p^2 + np - n^2p^2 = npq. \tag{10}$$

We leave the reader to show that if $\tilde{r}$ is $P(\mu)$ then

$$\mathscr{D}^2(\tilde{r}) = \mu. \tag{11}$$

Chebychev's inequality

The major use of Chebychev's inequality is in deriving other theoretical results, particularly the law of large numbers (§3.6). But in the form given in the corollary it does immediately show that a random variable can only rarely depart substantially from its mean, the actual departure depending on the standard deviation. For example, with $B(100, \frac{1}{2})$, which might arise with 100 tosses of a newly minted coin, the mean is $100 \times \frac{1}{2} = 50$ and the standard deviation is $\sqrt{(100 \times \frac{1}{2} \times \frac{1}{2})} = 5$ (equation (10)). Hence, with $c = 2$ in (8), the probability of either fewer than 40 or more than 60 heads is not more than 1/4. Actually the inequality, valid for *any* distribution with a variance, is too coarse for much use with those distributions which occur in practice. With a large class of distributions the right-hand side of (8) can be considerably reduced: for example with $c = 2$ we often have 1/20 instead of 1/4. In the binomial the probability of the event mentioned is about 1/20 (cf. §2.5).

Other features of distributions

Other features besides measures of location and spread are sometimes used. One particularly important one is the *coefficient of variation*, defined as the ratio of the standard deviation to the mean, σ/μ. This is only used when $\mu > 0$ and usually only for positive random variables. The idea behind its use is that a deviation of amount σ is less important when μ is large than when μ is small: a deviation of an inch in a mile is usually trivial compared with a deviation of an inch in a foot. The coefficient of variation of $\Gamma(n, \lambda)$ is $n^{-\frac{1}{2}}$, which does not depend on λ. It is quite common to find in practice that the coefficient of variation stays fairly constant even when μ and σ vary: for example, with wheat yields the mean changes from season to season and variety to variety but the coefficient remains at about 10%. It is commonly multiplied by 100 and expressed as a percentage. It is not affected by a change of units of the random variable. No standard notation is available; we shall use $\mathscr{V}(\tilde{x})$.

An important group of numbers associated with a distribution is the moments. The *n-th moment about the origin*, μ'_n, is defined as

$$\int x^n f(x)\,dx \quad \text{or} \quad \sum_r r^n p_r, \tag{12}$$

provided that

$$\int |x|^n f(x)\,dx \quad \text{or} \quad \sum_r |r|^n p_r \tag{13}$$

are finite. The latter is called the *n-th absolute moment about the origin*, ν'_n. If $\tilde{x}$ is a random variable with the corresponding distribution, we have

$$\mu'_n = \mathscr{E}(\tilde{x}^n), \quad \nu'_n = \mathscr{E}(|\tilde{x}|^n). \tag{14}$$

μ'_1 is the mean, also denoted by μ. The corresponding *moments about the mean* are similarly defined and the same notation is used without the primes. Thus

$$\mu_n = \int (x-\mu)^n f(x)\,dx = \mathscr{E}[(\tilde{x}-\mu)^n], \tag{15}$$

provided that the corresponding absolute moment,

$$\nu_n = \mathscr{E}(|\tilde{x}-\mu|^n),$$

is finite. μ_2 is the variance. Other moments are sometimes used, but not in this book. In modern work the moments are largely replaced by cumulants, to be introduced later (§2.6).

Histograms

In discussing in chapter 1 the relationship between the axioms and the real world, the concept of a limiting frequency observed in practice was intimately related to the notion of probability. We now introduce some further *empirical* material which corresponds to the *mathematical* concept of a density. In §1.3 we discussed the idea of random sampling from an infinite population and remarked that if each member of the population did, or did not, possess a characteristic then the sampling would be a sequence of random trials with constant probability of success (equal to the proportion of the population possessing the characteristic); and such trials exhibit the limiting success-ratio property. Consider again random sampling from a population but suppose that associated with each individual is an integer (or generally some discrete measurement), not the same for all individuals. In a sample of size n suppose m_r individuals have the associated integer r. Then if 'having the value r' is thought of as the event, A_r, a success, it is clear that '$\lim_{n\to\infty}$' m_r/n is $p_r = p(A_r)$. But (§2.1) $p(A_r)$ is the definition of the density of the random variable which is this measured integer. Consequently '$\lim_{n\to\infty}$' m_r/n is the real world counterpart of the density.

If the measurement, now x say, is continuous, a similar correspondence can be established by *grouping* the measurements. A measurement x is said to be grouped in intervals of width h if x is replaced by the nearest value in the series $\{c+nh\}$ ($n = 0, \pm 1, \pm 2, \ldots$): c is usually zero. Then if A_r is the event that the grouped value is $c+rh$, '$\lim_{n\to\infty}$' m_r/n is $p(A_r)$ which is approximately equal to $f(c+rh)h$, where $f(x)$ is the mathematical density. Hence '$\lim_{n\to\infty}$' m_r/nh corresponds to the density. In the discrete case with integers, h was one. The discussion may be summarized by saying that with a large n, $\{m_r/nh\}$ is the empirical counterpart of the density. Its graph against r is a

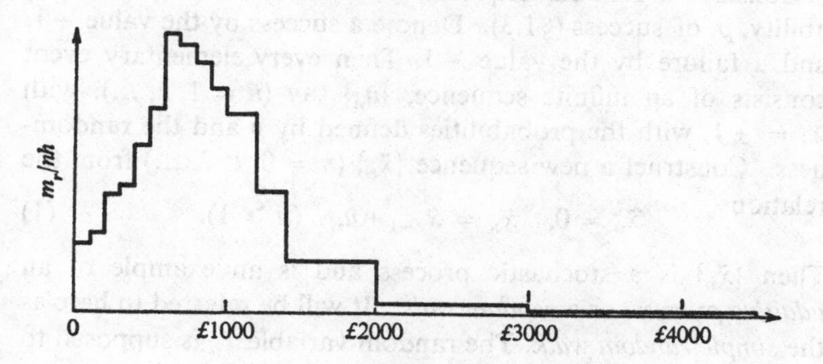

Fig. 2.4.1. Histogram of incomes.

histogram. The features of mathematical distributions (or densities) defined in this section may be similarly defined for histograms. For example, the mean is

$$\sum_r (c+rh)m_r/n \quad (= \sum_r rm_r/n \quad \text{if} \quad c = 0, h = 1)$$

corresponding to $\sum_r rp_r$. Such a feature is called a statistic. A *statistic* is a function of the sample values (cf. §5.5).

The mean has such attractive mathematical properties that within the mathematics it is almost the only measure of location used. But in descriptive statistics, where we are merely attempting to describe the data, the other measures of location are often no less useful than the mean. For example, the histogram† of the distribution of income in a population will often have the form sketched in fig. 2.4.1. The mean income will be affected by the few very rich people. The median income might be a better descriptive statistic for comparing two groups because it will not be so seriously influenced by the very rich. The mode of the income distribution has the same property and is the income-group of largest size in the population. Note that it will depend on the grouping used. A physical example that we shall meet later is the distribution of particle size (example 2 in §3.5).

2.5. The simple random walk

Consider a random sequence of trials with constant probability, p, of success (§1.3). Denote a success by the value $+1$, and a failure by the value -1. Then every elementary event consists of an infinite sequence, $\{\tilde{u}_n\}$ say ($n = 1, 2, \ldots$), with $\tilde{u}_n = \pm 1$, with the probabilities defined by p and the randomness. Construct a new sequence $\{\tilde{x}_n\}$ ($n = 0, 1, 2, \ldots$) from the relations

$$\tilde{x}_0 = 0, \quad \tilde{x}_n = \tilde{x}_{n-1} + \tilde{u}_n \quad (n \geqslant 1). \tag{1}$$

Then $\{\tilde{x}_n\}$ is a stochastic process and is an example of an *additive process*, or a *random walk*. It will be referred to here as the *simple random walk*. The random variable $\tilde{x}_n$ is supposed to represent the position after n steps of a particle which pursues a 'walk' on the integer points of the real line, starting from the origin and at each step moving to either of the neighbouring integer points according to the outcome of the random sequence. Our first theorem gives the density of $\tilde{x}_n$, for fixed n.

Theorem 1. *In a simple random walk $p(s \mid n)$, the probability that $\tilde{x}_n = s$, is equal to*

$$\binom{n}{\frac{1}{2}(n+s)} p^{\frac{1}{2}(n+s)} q^{\frac{1}{2}(n-s)}, \tag{2}$$

provided $\frac{1}{2}(n+s)$ is an integer in the interval $(0, n)$: otherwise it is zero.

Let r the number of $+1$'s, that is successes, in the first

† Notice that the groups are not equal. There are relatively few people with large incomes, so that h is increased for large incomes to enable a sensible number to appear in each group. Here $n = 27{,}000$.

n trials. Then $\tilde{x}_n$ will equal s iff the number of $+1$'s exceeds the number of -1's by s (s may be negative). Hence

$$s = r-(n-r) = 2r-n, \quad \text{or} \quad r = \tfrac{1}{2}(n+s).$$

But $\tilde{r}$ has a binomial distribution $B(n, p)$ with density given by equation 2.1.1, and (2) is merely equation 2.1.1 with $r = \frac{1}{2}(n+s)$.

Corollary. $\mathscr{E}(\tilde{x}_n) = n(p-q)$, $\mathscr{D}^2(\tilde{x}_n) = 4npq$.

From equation 2.1.9 $\mathscr{E}(\tilde{r}) = np$, and from equation 2.4.10 $\mathscr{D}^2(\tilde{r}) = npq$. Hence by equation 2.4.1

$$\mathscr{E}(\tilde{x}_n) = \mathscr{E}(2\tilde{r}-n) = 2\mathscr{E}(\tilde{r})-n = n(p-q),$$

and by equation 2.4.4

$$\mathscr{D}^2(\tilde{x}_n) = \mathscr{D}^2(2\tilde{r}-n) = 4\mathscr{D}^2(\tilde{r}) = 4npq.$$

An alternative derivation of (2) is important because it illustrates a method of wide applicability. This method consists in relating the position of a process now with the positions it could have just had. If $\tilde{x}_n$ is thought of as the position now, then $\tilde{x}_{n-1}$ could have been $\tilde{x}_n+1$, or $\tilde{x}_n-1$. Hence

$$p(s\,|\,n) = p(s-1\,|\,n-1)\,p+p(s+1\,|\,n-1)\,q, \tag{3}$$

from the generalized addition law (theorem 1.4.4) with the following correspondence between events,

$$A_k\colon \tilde{x}_{n-1} = k \quad \text{and} \quad B\colon \tilde{x}_n = s.$$

The $\{A_k\}$ $(k = 0, \pm 1, \pm 2, \ldots)$ are certainly exclusive and exhaustive, $p(A_k) = p(k\,|\,n-1)$, and the only conditional probabilities which do not vanish are $p(B\,|\,A_{s-1})$ and $p(B\,|\,A_{s+1})$ of values p and q respectively. Equation (2) obviously satisfies (3) with $p(0\,|\,0) = 1$ as the boundary condition, and can easily be shown to be the only solution with this boundary condition.

We are interested in the behaviour of (2) as $n \to \infty$. To study this we introduce a time variable (not a random variable) and suppose the successive steps in the walk take place at time instants δt apart. We consider the position after fixed time t when $n = t/\delta t$ steps will have occurred: then if $\delta t \to 0$, $n \to \infty$, and we shall, in the limit, be discussing continuous time. The position of the walk can also be made continuous in the limit by replacing the steps, ± 1, by steps, $\pm\,\delta x$, and allowing δx to tend to zero.

Our object is to pass from the difference equation (3) to a differential equation by means of these changes of variable.

Now if the density (2) is to tend to a limit as $n \to \infty$ the mean and variance should stay finite: provided they do, we know from Chebychev's inequality that the probability cannot be distributed too far from the mean. Also the variance must not tend to zero otherwise the distribution will have no scatter and be useless. The mean and variance of $\tilde{x}_n^*$, the position in the new units, will be, from the corollary, on change of units,

$$\mathscr{E}(\tilde{x}_n^*) = t(p-q)\,\delta x/\delta t, \quad \mathscr{D}^2(\tilde{x}_n^*) = 4pqt(\delta x)^2/\delta t. \tag{4}$$

To take care of the variance allow δx and δt to tend to zero so that

$$4pq(\delta x)^2/\delta t \to \sigma^2 > 0, \tag{5}$$

and to take care of the mean subtract from each step, $\pm\,\delta x$, the value $(p-q)\,\delta x$ and consider $\tilde{x}_n' = \tilde{x}_n^* - t(p-q)\,\delta x/\delta t$, with zero mean. Since we are changing from a discrete to a continuous time variable,† replace s in (3) by x and let $p(x\,|\,t) = p(\tilde{x}_n' = x)$. Then (3) becomes

$$p(x\,|\,t) = p(x-2q\,\delta x\,|\,t-\delta t)p + p(x+2p\,\delta x\,|\,t-\delta t)\,q. \tag{6}$$

(For example, if the particle is at $x-2q\,\delta x$ at time $t-\delta t$, the step $+\,\delta x$, of probability p will, when corrected by $(p-q)\,\delta x$, take the particle to $x-2q\,\delta x+\delta x-(p-q)\,\delta x = x$ at time t.) Subtract $p(x\,|\,t-\delta t)$ from both sides of (6) and divide by δt. Then

$$\frac{p(x\,|\,t)-p(x\,|\,t-\delta t)}{\delta t}$$

$$= \frac{2pq(\delta x)^2}{\delta t}\left\{\frac{p(x-2q\,\delta x\,|\,t-\delta t)p - p(x\,|\,t-\delta t) + p(x+2p\,\delta x\,|\,t-\delta t)q}{2pq(\delta x)^2}\right\}.$$

Allow δt and δx to tend to zero according to (5) and we have

$$\frac{\partial p(x\,|\,t)}{\partial t} = \tfrac{1}{2}\sigma^2\,\frac{\partial^2 p(x\,|\,t)}{\partial x^2}, \tag{7}$$

since the part in braces is a second difference at unequal intervals. Equation (7) is a famous partial differential equation, known as

† This point will be discussed more fully below, in order to avoid interrupting the proof.

the Fokker–Planck, *diffusion* or heat equation: $\frac{1}{2}\sigma^2$ is called the *diffusion coefficient*. The solution to (7) with the boundary condition that the particle at $t = 0$ was at $x = 0$ is easily found by the usual methods, but it is enough for us to observe that

$$p(x\,|\,t) = (2\pi t\sigma^2)^{-\frac{1}{2}}\exp\left(-x^2/2\sigma^2 t\right) \tag{8}$$

satisfies the equation and the boundary condition as $t \to 0$. To see that it satisfies the equation it is enough to carry out the differentiation. The boundary condition is most easily discussed after we have studied (8); which we now proceed to do.

It is easy to see that $p(x\,|\,t)$ satisfies the conditions (2.2.4) for a (continuous) density: for it is obviously positive, and the integral gives, on substituting $z = x(t\sigma^2)^{-\frac{1}{2}}$, the integral $\int_{-\infty}^{\infty} e^{-\frac{1}{2}z^2}\,dz/\sqrt{(2\pi)}$ which is known to be unity (see below). The mean of the density is obviously zero since (8) is symmetric about zero. The variance is therefore $\int x^2 p(x\,|\,t)\,dx$ which gives $\sigma^2 t\int_{-\infty}^{\infty} z^2 e^{-\frac{1}{2}z^2}\,dz/\sqrt{(2\pi)}$ with the same substitution. An integration by parts reduces the integral to the one just considered and hence the value is $\sigma^2 t$. As $t \to 0$, therefore, the mean and variance tend to zero and the random variable tends to be at the origin with probability one. This establishes the correctness of the boundary condition. There is no loss in generality, when $t > 0$, in supposing $t = 1$. Equation (8) is said to be a *normal* (or Gaussian, or Laplacian) density and a random variable with that density is said to have a normal distribution with mean zero and variance σ^2. This is written $N(0, \sigma^2)$. Still with $t = 1$ we can state the main result we have proved as

Theorem 2. *As $n \to \infty$ the distribution of $[\tilde{x}_n - n(p-q)]/(4npq)^{\frac{1}{2}}$ tends to the normal distribution $N(0, 1)$.*

We have shown that the distribution of $[\tilde{x}_n^* - t(p-q)\,\delta x/\delta t]$ tends to $N(0, \sigma^2 t)$, for fixed t, as $\delta t \to 0$, $\delta x \to 0$ and therefore $n = (t/\delta t) \to \infty$. Dividing by the standard deviation $\sigma t^{\frac{1}{2}}$ and remembering (5) and that $\tilde{x}_n^* = \tilde{x}_n\,\delta x$ we see that

$$\frac{\tilde{x}_n\,\delta x - t(p-q)\,\delta x/\delta t}{[4tpq(\delta x)^2/\delta t]^{\frac{1}{2}}}$$

tends to $N(0, 1)$ which, with $n = t/\delta t$, is the required result.

Corollary. (*De Moivre's theorem.*) *If* $\tilde{r}$ *is* $B(n, p)$ *then the distribution of* $[\tilde{r}-np]/(npq)^{\frac{1}{2}}$ *tends to the normal distribution* $N(0, 1)$ *as* $n \to \infty$.

This is immediate since $\tilde{x}_n = 2\tilde{r}-n$.

If $\tilde{x}$ is $N(0, \sigma^2)$ then $\tilde{x}+\mu$ clearly has density

$$(2\pi\sigma^2)^{-\frac{1}{2}} \exp[-(x-\mu)^2/2\sigma^2] \tag{9}$$

with mean μ. This is a normal density $N(\mu, \sigma^2)$. μ and σ^2 are called the *parameters* of the normal distribution. The distribution function is

$$(2\pi\sigma^2)^{-\frac{1}{2}} \int_{-\infty}^{x} \exp[-(y-\mu)^2/2\sigma^2]\, dy \tag{10}$$

which, on substituting $(y-\mu)/\sigma = z$, is

$$(2\pi)^{-\frac{1}{2}} \int_{-\infty}^{(x-\mu)/\sigma} e^{-\frac{1}{2}z^2}\, dz. \tag{11}$$

The function

$$(2\pi)^{-\frac{1}{2}} \int_{-\infty}^{x} e^{-\frac{1}{2}z^2}\, dz \tag{12}$$

is usually denoted by $\Phi(x)$, and its derivative $(2\pi)^{-\frac{1}{2}} e^{-\frac{1}{2}x^2}$ by $\phi(x)$. From (11) $\Phi([x-\mu]/\sigma)$ is the distribution function of $N(\mu, \sigma^2)$. It follows that if $\tilde{x}$ is $N(\mu, \sigma^2)$ then $(\tilde{x}-\mu)/\sigma$ is $N(0, 1)$. A random variable which is $N(0, 1)$ is said to be a *standardized normal variable*.

Examples of random walks

The simple random walk probably first arose in games of chance. Consider a sequence of plays of a game: if you win any play you receive 1 unit; if you lose you pay out 1 unit, that is receive -1. The successive plays are presumably independent (unless you change strategy according to the results of earlier plays) with constant probability of winning; hence the $\{\tilde{u}_n\}$ above correspond to the receipts in the plays and $\tilde{x}_n$ is one player's capital after n plays, starting from zero. The process is also a crude model for the diffusion of a particle in a fluid. This is the one-dimensional form in which the particle undergoes bombardment either from the right or left causing it to move one step. Another example occurs in industrial inspection:

mass-produced articles are randomly sampled giving rise to the binomial distribution (§2.1). If one is found satisfactory then a 'score' of $+1$ is given, otherwise the score is -1. The total score obviously forms a simple random walk.

The adjective 'simple' refers to the case where $\tilde{u}_n$ is either plus or minus one. If the $\tilde{u}$'s have a more general (common) distribution, and are independent, then we have a random walk or an additive process. The reason for the latter name is that the process $\{\tilde{x}_n\}$ is formed by *adding* a random variable $\tilde{u}_n$, at any stage of the process, $\tilde{x}_{n-1}$, to obtain the next stage $\tilde{x}_n$. In many applications of random walks there exist *barriers*, in the sense that once the walk reaches one of them it either stops or is prevented from proceeding beyond them. For example, games of chance may stop when either of the players become bankrupt. Such walks will be discussed in §§4.5 and 4.6.

Normal approximation to the binomial

The binomial density, or (2), is tedious to calculate and if tabulated requires tables of triple entry, for n, r and p (Harvard Computation Laboratory, 1955; Romig, 1953). It is therefore natural to search for approximations. We already have one in the Poisson distribution (§2.3) valid as $n \to \infty$, $p \to 0$, but it would be desirable to have one for any p and large n; for this is when the calculation of the binomial density becomes tedious. De Moivre's theorem gives us such an approximation. We illustrate the use of this theorem by calculating the probability of 30 or more heads in 50 tosses of a fair coin—not too easy a task using the binomial distribution. The number of heads, $\tilde{r}$, is $B(50, \frac{1}{2})$ and therefore has mean 25 and standard deviation $\sqrt{(50 \times \frac{1}{2} \times \frac{1}{2})} = 3{\cdot}54$. Consequently the density of $(\tilde{r}-25)/3{\cdot}54$ is approximately $N(0, 1)$. Since $\tilde{r} \geqslant 30$ corresponds to $(\tilde{r}-25)/3{\cdot}54 \geqslant 1{\cdot}41$ the required probability is approximately $1-\Phi(1{\cdot}41)$, where $\Phi(x)$ is the distribution function of an $N(0, 1)$ variable. Tables of $\Phi(x)$ give the value 0·079. The approximation can be improved in a way which is best explained by a figure. Fig. 2.5.1 shows the distribution function of $(\tilde{r}-25)/3{\cdot}54$, in the neighbourhood of $r = 30$, plotted on the same scale as that of $N(0, 1)$. The alternative scale, approximating $B(50, \frac{1}{2})$ by $N(25, 12\frac{1}{2})$, is also shown. It is clear from the figure that the probability of 30 or more (the distance the point A is below one) is more nearly approximated by taking the corresponding normal expression at $r = 29\frac{1}{2}$, rather than 30. We then obtain $1-\Phi(1\cdot27) = 0\cdot102$. The exact value obtained from the cumbersome binomial tables is 0·1013. Sometimes the correction of $\frac{1}{2}$ has to be added instead of subtracted: which is appropriate will be clear from a sketch. It is often called a *continuity* correction.

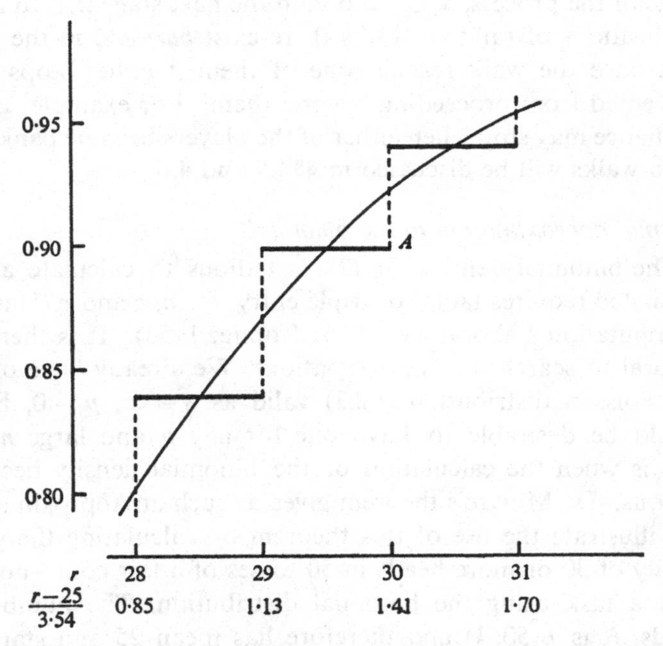

Fig. 2.5.1. Distribution functions for $B(50, \frac{1}{2})$ and the normal approximation $N(25, 12\frac{1}{2})$.

The great merit of the approximation is that it enables a table of triple entry to be replaced by one of single entry. Every binomial distribution can be reduced in the way just exemplified to the normal form. $\Phi(x)$ is not expressible in terms of elementary functions but has to be regarded as a new function: to anyone working with random phenomena it is as important as the sine function is to a surveyor. The normal distribution occurs in many studies and is used more often than any other distribution. It is extensively tabulated, the most comprehensive being that of the National Bureau of Standards (1953): the simpler table of Lindley and Miller (1961) will be enough for our needs. Since the distribution function of $N(\mu, \sigma)$ is $\Phi[(x-\mu)/\sigma]$ it follows that once the standardized normal distribution is tabulated, all normal distributions are effectively tabulated. This is why the approximation is so useful. Since the table has to be entered at $(x-\mu)/\sigma$, or in the binomial case $(r-np)/(npq)^{\frac{1}{2}}$, the relevant value for the probability calculation is the number of standard deviations that x (or r) is from the mean. In the example, 30 is 1·41 standard deviations from the mean. Notice that since the normal density is symmetric about the mean, $\Phi(x) = 1-\Phi(-x)$. Hence $\Phi(x)$ need only be tabulated for positive argument. It is worth knowing a few values of $\Phi(x)$ just as one knows a few values of the trigonometric functions. Three useful ones are

$$\Phi(1{\cdot}96) = 0{\cdot}975, \quad \Phi(2{\cdot}58) = 0{\cdot}995, \quad \Phi(3{\cdot}29) = 0{\cdot}9995. \tag{13}$$

It is instructive to compare these values with the limits obtained by Chebychev's inequality. If $\tilde{x}$ is a normal random variable

$$\begin{aligned} p(|\tilde{x}-\mu| \geqslant c\sigma) &= p(\tilde{x}-\mu \geqslant c\sigma)+p(\tilde{x}-\mu \leqslant -c\sigma) \\ &= 1-\Phi(c)+\Phi(-c) = 2\Phi(-c). \end{aligned} \tag{14}$$

With $c = 1{\cdot}96$ this gives a probability of twice $(1-0{\cdot}975)$, i.e. 0·05, or 1 in 20. Chebychev's inequality says that the probability is less than $(1{\cdot}96)^{-2}$, about 1 in 4. Thus the inequality is very coarse. The three numerical values quoted above are equivalent to saying 'the probability that a normal random variable departs from its mean by more than 1·96 (2·58 or 3·29) standard deviations is 1 in 20 (100 or 1000)'.

Limit of the distribution function

In the main theorem the density for a discrete distribution (the binomial) has been approximated by the density for a continuous distribution. In applications it is usually the distribution function that is needed (as in the numerical example just given) and it is therefore important to notice that theorem 2 persists with distribution function for density. This is easily seen by summing (6) over all x less than some value, which shows that the (discrete) distribution function also satisfies (6) and hence, in the limit, (7). In solving (7) the boundary conditions are different, for at $t = 0$ the distribution function is zero for $x < 0$ and 1 for $x \geqslant 0$. The solution with this modification is $\Phi(x/\sigma t^{\frac{1}{2}})$, agreeing with the density. It is also easier to understand how the discrete probability distribution passes into the continuous one, if the distribution function replaces the density. The 'jumps' in the former become smaller, and because of the scaling (by δx) become closer together: consequently they are 'smoothed out' and pass to a continuous limit. Fig. 2.5.1 will help to clarify this.

Diffusion with drift

The limiting *diffusion* process, satisfying (7), has always zero mean (since (8) has zero mean). A *drift* of amount μ per unit time can be introduced on replacing x by $x+\mu t$. The usual methods for change of variable give

$$\frac{\partial p(x\,|\,t)}{\partial t}+\mu\frac{\partial p(x\,|\,t)}{\partial x} = \tfrac{1}{2}\sigma^2\frac{\partial^2 p(x\,|\,t)}{\partial x^2}, \tag{15}$$

the general diffusion equation. The same change of variable shows that the solution is that the distribution, at time t, is $N(\mu t, \sigma^2 t)$.

Evaluation of an integral

The result
$$\int_{-\infty}^{\infty} e^{-\frac{1}{2}z^2}\,dz = \sqrt{(2\pi)}, \tag{16}$$

used above, is easily proved as follows. The double integral of $e^{-\frac{1}{2}(x^2+y^2)}$ over a square of side $2R$ with centre at the origin and sides parallel to the x- and y-axes is obviously, since the integrand is positive, bounded below and above by the same integral over circles of radii R and $\sqrt{2}R$ respectively with centres at the origin. On changing to polar co-ordinates the former integral becomes†

$$\int_0^R dr \int_0^{2\pi} d\theta\, e^{-\frac{1}{2}r^2} r,$$

since the element of area $dx\,dy$ becomes $r\,dr\,d\theta$. As $R \to \infty$ this gives

$$2\pi \int_0^\infty e^{-\frac{1}{2}r^2} d(\tfrac{1}{2}r^2) = 2\pi.$$

Similarly, the upper bound gives the same value; hence

$$\int_{-\infty}^{\infty} dx \int_{-\infty}^{\infty} dy\, e^{-\frac{1}{2}(x^2+y^2)} = \left[\int_{-\infty}^{\infty} e^{-\frac{1}{2}z^2} dz\right]^2 = 2\pi.$$

Notice that the substitution $\frac{1}{2}z^2 = t$, gives

$$\int_0^\infty e^{-t} t^{-\frac{1}{2}} dt = \sqrt{\pi},$$

previously defined (§2.3) as $(-\frac{1}{2})!$.

Central limit theorem

An alternative form of theorem 2 is important for later extensions. From (1) it follows that $\tilde{x}_n = \sum_{r=1}^{n} \tilde{u}_r$ and hence the theorem says that the sum of the $\{\tilde{u}_n\}$ is, as their number increases, approximately normally distributed. We shall see later that this is true for cases where the $\{\tilde{u}_n\}$ are fairly general random variables and not merely ones which assume only the values ± 1. This result is the central limit theorem (3.6.1) and is one of the reasons for the importance of the normal distribution.

2.6. Generating functions

Let $\{p_r\}$ be the density of a non-negative, integer-valued random variable. The function

$$\Pi(x) = \sum_{r=0}^{\infty} p_r x^r \tag{1}$$

is called the *probability generating function* of the distribution (or density). The argument x may be complex, and (1) certainly

† In this book multiple integrals will have the variable of integration placed next to the sign of integration so that no confusion arises over the range of each variable.

converges for $|x| \leqslant 1$ since Σp_r does, and so is at least defined within and on the unit circle. If the random variable can assume negative values the probability generating function can still be defined by (1), with summation now from minus to plus infinity, but may only exist for $|x| = 1$. In what follows we suppose it does exist for other values of x, and use the doubly infinite sum.

The derivatives of $\Pi(x)$ at $x = 0$ involve the density $\{p_r\}$. The derivatives at $x = 1$ involve the moments (§2.4). For example,

$$\left(\frac{d\Pi}{dx}\right)_{x=1} = \mu \quad \text{and} \quad \left(\frac{d^2\Pi}{dx^2}\right)_{x=1} = \mathscr{E}[\tilde{r}(\tilde{r}-1)] = \sigma^2+\mu^2-\mu. \quad (2)$$

The second of these follows from theorem 2.4.2. The moments may be generated directly by putting $x = e^z$ in (1) and defining†

$$\Phi(z) = \sum_{r=-\infty}^{\infty} p_r e^{rz}. \quad (3)$$

$\Phi(z)$ is called the *moment generating function*. If $\Pi(x)$ exists in a neighbourhood of $|x| = 1$, $\Phi(z)$ exists in a neighbourhood of the real part of $z = 0$. We have, on expanding the exponential function,

$$\Phi(z) = \sum_{n=0}^{\infty} \mu_n' z^n/n! \quad \text{and} \quad (d^n\Phi(z)/dz^n)_{z=0} = \mu_n'. \quad (4)$$

(It will agree with the definition of μ_n' (equation 2.4.14) to put $\mu_0' = 1$.) The more useful moments about the mean can be similarly obtained by using the function $\Phi(z)e^{-\mu z}$. Notice that $\Phi(z) = \mathscr{E}(e^{\tilde{r}z})$.

The moment generating function may only exist when $z = it$, t real, corresponding to $|x| = 1$, when the above results may be invalid. In theoretical work, in order to avoid unnecessary assumptions about the distribution, it is therefore convenient to use the *characteristic function* defined as

$$\Psi(t) = \Phi(it), \quad (5)$$

where t is real and $i = \sqrt{-1}$. $\Psi(t)$ exists for all real t.

We shall see later (theorem 3.5.4) that $\Phi(z)$, and therefore $\Psi(t)$, has some useful multiplicative properties. It is therefore convenient to introduce the function

$$K(z) = \ln \Phi(z), \quad (6)$$

† The notation should not be confused with that for the normal distribution function.

which will have a power series expansion, if $\Phi(z)$ does, equal to

$$K(z) = \sum_{n=1}^{\infty} \kappa_n z^n/n!. \tag{7}$$

(Notice the $n!$ which occurs in the expansion of $\Phi(z)$, equation (4), and $K(z)$.) The $\{\kappa_n\}$ are called the *cumulants* of the distribution and $K(z)$ is called the *cumulant generating function*. The cumulants often replace the moments in modern work: they are related to them by the result

$$\ln\left(\sum_{n=0}^{\infty} \mu_n' z^n/n!\right) = \sum_{n=1}^{\infty} \kappa_n z^n/n!.$$

In particular

$$\kappa_1 = \mu, \quad \kappa_2 = \sigma^2, \quad \kappa_3 = \mu_3, \quad \kappa_4 = \mu_4 - 3\sigma^4.$$

The infinite expansions in (4) and (7) are not always possible because the moments or cumulants may not exist. However, if μ_n (or μ_n') exists for some n (and therefore, it is easy to see, for all smaller n) we may write

$$\Phi(z) = \sum_{j=0}^{n} \mu_j' z^j/j! + R(z), \tag{8}$$

where $R(z)$ is a remainder term such that $R(z)/z^n \to 0$ as $z \to 0$, which is written $R(z) = o(z^n)$. Thus $\Phi(z)$ generates the moments as far as they exist. Similarly,

$$K(z) = \sum_{j=1}^{n} \kappa_j z^j/j! + R^*(z). \tag{9}$$

With continuous distributions the probability generating function is no longer used, but the definition of the moment generating function as $\Phi(z) = \mathscr{E}(e^{\tilde{x}z})$ applies to this case also. If $f(x)$ is a continuous density, then

$$\Phi(z) = \int f(x) e^{xz} dx. \tag{10}$$

It similarly generates the moments, certainly exists when $z = it$, and gives the characteristic function and the cumulants as in the discrete case.

The uses of generating functions

Probability distributions can be described in several ways, through the density function or the distribution function, for example. Which way is chosen depends on the problem and the properties of the descriptive function. The generating functions are descriptions which have some particularly attractive and simple properties in a problem involving the sum of random variables; for example, the random walk in the form mentioned at the end of §2.5. This will be discussed in §3.6. Furthermore, unlike the measures introduced in §2.4, they describe the distribution completely; that is, given a generating function there is only one distribution with that as its generating function: they characterize the distribution. (Hence the name, characteristic function.) The moments do not characterize the distribution.

We have given definitions of the generating functions in terms of the density function; formulae which go in the reverse way, from the generating function to the density (or distribution) function are called *inversion* formulae. We shall not prove that characteristic functions characterize the distribution, nor shall we give the inversion formulae, which are not often useful. It will be enough for our purpose to recognize the generating functions of the distributions we shall meet. An example is given after (20) below. Inversion formulae are given in more advanced books, for example Loève (1960). In the discrete case the expansion of $\Pi(x)$ as a power series will give the density: but this is not available in the continuous case.

Some readers may be familiar with generating functions under other names. For example, the moment generating function in the continuous case (equation (10)) is the two-sided Laplace transform of $f(x)$. The characteristic function is the two-sided Fourier transform. If the random variables are positive the transforms are one-sided. It might appear that much use could be made of the extensive theory of Laplace and Fourier transforms, but often the particular property of $f(x)$ of being non-negative makes the transforms more special than they are in the general theory, and therefore they have extra properties not always discussed in the literature on transforms.

Example 1. The Poisson distribution. The probability generating function is

$$\Pi(x) = \sum_{r=0}^{\infty} e^{-\mu}\mu^r x^r/r! = \exp[\mu(x-1)] \tag{11}$$

which exists for all x. The derivative at $x = 1$ is clearly μ, the mean, agreeing with (2). The moment generating function is obtained by putting $x = e^z$ in (11), and is

$$\Phi(z) = \exp[\mu(e^z - 1)]. \tag{12}$$

About the mean the function is $\Phi(z)e^{-\mu z} = \exp[\mu(e^z - 1 - z)]$. The expansion of this begins $1 + \mu z^2/2! + \ldots$ so that the variance $\mu_2 = \sigma^2$ is μ. Finally $K(z) = \mu(e^z - 1)$ so that all the cumulants of the Poisson distribution are equal to μ.

Example 2. The normal distribution. The density is given by equation (2.5.9), so that the moment generating function is

$$\begin{aligned}\Phi(z) &= (2\pi\sigma^2)^{-\frac{1}{2}} \int_{-\infty}^{\infty} \exp[zx - (x-\mu)^2/2\sigma^2]\, dx \\ &= (2\pi\sigma^2)^{-\frac{1}{2}} \int_{-\infty}^{\infty} \exp[-\{(x - \mu - z\sigma^2)^2 \\ &\qquad\qquad - 2z\mu\sigma^2 - z^2\sigma^4\}/2\sigma^2]\, dx,\end{aligned}$$

on completing the square in x. The integral is reduced to the usual form (2.5.16) by the substitution $x - \mu - z\sigma = t$, so that

$$\Phi(z) = \exp[z\mu + \tfrac{1}{2}z^2\sigma^2]. \tag{13}$$

The derivative at the origin is μ, which is already known to be the mean, so that the moment generating function about the mean is

$$\Phi(z)e^{-\mu z} = \exp[\tfrac{1}{2}z^2\sigma^2], \tag{14}$$

from which it follows that

$$\begin{aligned}\mu_{2r+1} = 0 \quad \text{and} \quad \mu_{2r} &= \frac{\sigma^{2r}(2r)!}{2^r r!} \\ &= (2r-1)(2r-3)\ldots 3\sigma^{2r}.\end{aligned} \tag{15}$$

The cumulant generating function is $z\mu + \frac{1}{2}z^2\sigma^2$ so that $\kappa_1 = \mu$, $\kappa_2 = \sigma^2$ and all the higher cumulants are zero.

Example 3. The Poisson process. (This example also serves as another illustration of the important probability technique used to derive equation 2.5.3.) The notation is that of §2.3. We derive an equation for $p_n(t)$ by relating the number of incidents up to time t, thought of as 'now', with the number up to a point just in the 'past', namely at $t - \delta t$. We saw that the Poisson process could be formulated by saying that the probability of one incident in an interval $(t-\delta t, t)$ is $\lambda\delta t$, and of two or more is small in comparison with δt, independent of the behaviour of the process prior to $t-\delta t$. Consequently there can only be n incidents up to t if either there were n up to $t-\delta t$ and none took place in $(t-\delta t, t)$ (the latter event has probability $1-\lambda\delta t$), or if there were $(n-1)$ up to $t-\delta t$ and one occurred in $(t-\delta t, t)$ (of probability $\lambda\delta t$). Hence by the generalized addition law (theorem 1.4.4), provided $n > 0$ and terms of smaller order than δt are neglected,

$$\begin{aligned} p_n(t) &= p_{n-1}(t-\delta t)\, p[n \text{ incidents in } (0, t) \,|\, (n-1) \text{ incidents in } (0, t-\delta t)] \\ &\quad + p_n(t-\delta t)\, p[n \text{ incidents in } (0, t) \,|\, n \text{ incidents in } (0, t-\delta t)] \\ &= p_{n-1}(t-\delta t)\,\lambda\,\delta t + p_n(t-\delta t)\,[1-\lambda\,\delta t]. \end{aligned} \tag{16}$$

The reader should make sure he has understood the argument used in deriving (16), since it will be used again (§4.1), and the use of the independence condition in the derivation. Equation (16) may be rewritten

$$[p_n(t)-p_n(t-\delta t)]/\delta t = \lambda[p_{n-1}(t-\delta t)-p_n(t-\delta t)],$$

which, on allowing δt to tend to zero, gives

$$dp_n(t)/dt = \lambda[p_{n-1}(t)-p_n(t)]. \tag{17}$$

This is a differential-difference equation for $p_n(t)$: that is, it involves differentials with respect to t and differences with respect to n. The boundary conditions are $p_0(0) = 1$, $p_n(0) = 0$ for $n > 0$. Equation (16), and hence (17), also holds for $n = 0$ provided $p_{-1}(t)$ is put equal to zero. The equations can be solved in succession starting with $n = 0$; instead we shall solve them more simply using probability generating functions. Let

$$\Pi(x, t) = \sum_{n=0}^{\infty} p_n(t)\, x^n, \tag{18}$$

the probability generating function for the number of incidents in $(0, t)$. Multiply (17) by x^n and sum the two sides over n. The result is

$$\partial\Pi(x, t)/\partial t = \lambda(x-1)\,\Pi(x, t). \tag{19}$$

As there are no differentials with respect to x in the equation, x may be regarded as constant; t is the variable and the equation is an ordinary (and not a partial) differential equation. We easily have

$$d\Pi/\Pi = \lambda(x-1)\,dt,$$

so that

$$\ln \Pi(x, t) = \lambda(x-1)t + f(x),$$

where $f(x)$ is an arbitrary function. The boundary condition at $t = 0$ gives $\Pi(x, 0) = 1$ so that $f(x) = 0$ and

$$\Pi(x, t) = \exp[\lambda t(x-1)]. \qquad (20)$$

A comparison of (20) with (11) and the fact that the probability generating function characterizes the distribution shows that $\{p_n(t)\}$ must be a Poisson distribution with parameter λt. This is theorem 2.3.3 again.

Suggestions for further reading

The two books on the calculus of probabilities mentioned in the suggestions in chapter 1 by Feller (1957) and Gnedenko (1962) cover material closely related to that of this chapter and the two succeeding. The Poisson process is discussed by Khintchine (1960), and in the introductory book on probability by Parzen (1962). A more advanced book on probability is Loève (1960).

A branch of the subject, not discussed here, is geometrical probability; see Kendall and Moran (1963).

Exercises

1. In an investigation of animal behaviour, rats have to choose between four similar doors, one of which is 'correct'. Correct choice is rewarded by food and incorrect choice is punished by an electric shock. If an incorrect choice is made, the rat is returned to the starting-point and chooses again, this continuing until the correct response is made. The random variable, X, is the serial number of the trial on which the correct response is first made, X thus taking values 1, 2, ...

Find the distribution and mean of X under the following different hypotheses:

(i) each door is equally likely to be chosen on each trial, and all trials are mutually independent;

(ii) at each trial, the rat chooses with equal probability between the doors that have not so far been tried, no choice ever being repeated;

(iii) the rat never chooses the same door on two successive trials, but otherwise chooses at random with equal probabilities. (Lond. B.Sc.)

2. In inspecting an industrial process, a sample of material is examined every hour and classified as 'satisfactory' (S), 'doubtful' (D), or 'unsatisfactory' (U). If S is observed, no further action is taken. If U is observed, corrective action is taken. If D is observed, an additional sample is taken at once and additional sampling continued until either S or U is obtained, when the appropriate action is taken. The probabilities of S, D, U are α, β, γ respectively ($\alpha+\beta+\gamma = 1$), independently for each sample.

Prove that the probability that corrective action is taken at a particular test period is $\gamma/(1-\beta)$.

Derive the distribution and mean of the number of times corrective action is taken in n inspection periods, α, β, γ remaining constant.

(Lond. B.Sc.)

3. In the game of 'Craps' (ex. 29 of chapter 1) find the expected number of throws in a single play.

4. A bag contains a very large number of balls of which a proportion p is red and a proportion $q = 1-p$ is white. In a psychological test the balls are drawn at random one at a time from the bag and after the first ball has been drawn a subject, who does not know the value of p, is asked to guess the colour of the second ball; when he has made his guess and the second ball has been taken from the bag and shown to the subject, he is asked to guess the colour of the third ball, and so on. Show that if the subject always guesses the same colour as the previous ball, the probability of any guess being correct is p^2+q^2; and that the same probability, given that the previous guess was correct, is $(p^3+q^3)/(p^2+q^2)$. Show that the latter exceeds the former and hence that the number of correct guesses is not binomially distributed. Can you suggest a better rule for making his guesses, independent of any knowledge of p?

5. The spores of a certain plant are arranged in sets of four in a linear chain (with three links). When the spores are projected from the plant, each link has a probability θ of breaking, independently for each link. For example, if all links break, four 'groups' each of one spore are obtained, whereas if no links break a single 'group' of four spores results. Prove that the expected number of 'groups' per set is $1+3\theta$.

A large number of 'groups' is collected. Show that the proportions of 'groups' having 1, 2, 3 and 4 spores are respectively

$$\frac{2\theta(1+\theta)}{1+3\theta}, \quad \frac{\theta(1-\theta)(2+\theta)}{1+3\theta}, \quad \frac{2\theta(1-\theta)^2}{1+3\theta}, \quad \text{and} \quad \frac{(1-\theta)^3}{1+3\theta}.$$

(Lond. B.Sc.)

6. A game between two players, A and B, consists in them playing alternately a machine until one of them scores a success on it. The first to score a success wins. Their probabilities of success on each play of the machine are respectively p_1 and p_2. Since B is a better player than A ($p_2 > p_1$) he allows A to have the first turn. All plays of the machine are independent.

Show that the game is fair iff

$$p_2 = p_1/(1-p_1).$$

Show that when A wins, the average number of plays he takes in which to win is

$$(p_1+p_2-p_1p_2)^{-1}.$$ (Wales Maths.)

7. A timber merchant knows from previous experience that, on average, 20% of the consignments of timber received by him come from plantations affected by disease, and 80% from disease-free plantations. On average 15% of the logs in consignments from diseased plantations have some defect, compared with 5% of logs from disease-free plantations. Logs are selected at random from a consignment and examined. The sampling continues until the first log with a defect is selected, when the number of defect-free logs sampled is noted. If the number of defect-free logs in the sample is 7, what is the probability that the consignment came from a diseased plantation? (Aberdeen Dip.)

8. It is assumed that the chance of a male birth is constantly equal to a known constant p and that the sex of any child is independent of that of any other child. A family is known to contain just r males. What are the posterior probabilities and means of the size, n, of this family if the distribution of family size in the population is:

$$\text{(i) } p(n) = \frac{2^n}{n!}e^{-2} \quad (n = 0, 1, 2, \ldots);$$

$$\text{(ii) } p(n) = 1/2^{n+1} \quad (n = 0, 1, 2, \ldots)?$$

(Lond. Dip.)

9. In a batch of N manufactured items, there are R defectives and $N-R$ non-defectives. Show that the distribution of $\tilde{r}$, the number of defectives in a random sample of n items from the batch, is given by the *hypergeometric* probability density function

$$p(\tilde{r} = r) = \binom{R}{r}\binom{N-R}{n-r}\bigg/\binom{N}{n} \quad (r = 0, 1, \ldots, n).$$

If R itself is a binomial random variable $\tilde{R}$, with parameter p and index N, use Bayes's theorem to show that, given $\tilde{r} = r$, the distribution of $\tilde{R}-r$ is independent of r.

10. A lake contains an unknown number N of fish. A sample of n fish is drawn from it and each specimen is marked without injuring it and replaced while still alive. A month later a second sample of k fish is drawn from the lake and is found to include exactly m marked specimens. Show that on the assumptions that the fish have become completely mixed and that none has died, entered or left the lake, or has been born during the interval, the probability of obtaining this result is given by

$$p_N = \binom{N-n}{k-m}\binom{n}{m}\bigg/\binom{N}{k}.$$

The size of the population is estimated† by choosing that value of N which makes p_N as great as possible. By considering the ratio $u_N = p_N/p_{N-1}$, or otherwise, show that the estimated number is approximately nk/m.

11. An urn contains n balls, each a different colour. Let r be any integer. Show that the probability that a sample of size r, drawn with replacement, will contain r_1 balls of colour 1, r_2 balls of colour 2, ..., r_n balls of colour n (where $r_1+r_2+\ldots+r_n = r$) is given by

$$\frac{1}{n^r}\frac{r!}{r_1!\,r_2!\ldots r_n!}.$$

12. A radioactive source, emitting α-particles in random directions (all equally likely) is held at unit distance from an infinite plane photographic plate. If an α-particle can be stopped by the plate only, what is the distribution of $1/x$ where x is the distance of the point of impact of a particle from the point of the plate nearest the source?

13. A patient with a needle 5 cm long in his chest is X-rayed. If the orientation of the needle is quite random, what is the distribution function of the length of the needle's shadow? What are the mean and variance of the shadow length if the needle has a uniform distribution of length between 3 and 7 cm? (A *uniform* distribution in (a, b) has constant density $(b-a)^{-1}$ in that interval.)

14. From a point on the circumference of a circle of radius a, a chord is drawn in a random direction (i.e. all directions are equally likely). Show that the mean of the length of the chord is $4a/\pi$, and that the variance of the length is $2a^2(1-8/\pi^2)$. Also, show that the chance is 1/3 that the length of the chord will exceed the length of the side of an equilateral triangle inscribed in the circle.

If the chord is drawn parallel to a given straight line, all distances from the centre of the circle being equally likely, show that the mean length of chord is $\frac{1}{2}\pi a$ and that the variance is $a^2(32-3\pi^2)/12$. Also show that the chance is now 1/2 that the length will exceed the side of the equilateral triangle.

15. A point P is taken at random in a line AB, of length $2a$, all positions of the point being equally likely. Show that the mean value of the area of the rectangle $AP.PB$ is $2a^2/3$ and that the probability of the area exceeding $\frac{1}{2}a^2$ is $1/\sqrt{2}$.

16. Suppose the duration, t, in minutes of long-distance telephone calls made from a certain city is found to have a distribution function

$$\begin{aligned} F(t) &= 0 \quad \text{for} \quad t \leqslant 0 \\ &= 1-\tfrac{2}{3}e^{-\frac{1}{3}t}-\tfrac{1}{3}e^{-[\frac{1}{3}t]} \quad \text{for} \quad t > 0. \end{aligned}$$

($[\frac{1}{3}t]$ is the integral part of $\frac{1}{3}t$.)

(i) Sketch the distribution function.

† Notice that this is an inference problem. Estimation will be formally discussed in §§5.2 and 7.1.

(ii) Is the random variable t continuous, discrete, neither?

(iii) What is the probability that the duration in minutes of a long-distance call will be (*a*) more than 6, (*b*) less than 4, (*c*) equal to 3, (*d*) between 4 and 7?

(iv) What is the conditional probability that the duration of a call will be less than 9 minutes, given that it has lasted more than 5 minutes?

17. The probability density of the velocity, v, of a molecule with mass m in a gas at absolute temperature T is

$$\begin{aligned} g(v) &= \alpha v^2 e^{-\beta v^2}, \quad \text{for} \quad v \geqslant 0, \\ &= 0, \quad \text{for} \quad v < 0 \end{aligned}$$

(Maxwell–Boltzmann law): where $\beta = m/2kT$, k is Boltzmann's constant and α is a constant so chosen that

$$\int_0^\infty g(v)\,dv = 1.$$

Find the mean and variance of v, and also the mean and variance of the kinetic energy $E = \frac{1}{2}mv^2$. Show that the mean and variance of $E/(\frac{1}{2}kT)$ are those of a $\Gamma(\frac{3}{2}, \frac{1}{2})$ variable.

18. Certain metal bars are such that the probability of a flaw in any section of length δx is, for sufficiently small δx, equal to $\lambda\,\delta x$, the probabilities associated with different sections being independent. Prove that the probability that a rod of length l is without a flaw is $e^{-\lambda l}$.

If a rod has no flaw its strength is S_1 and if it contains one or more flaws its strength is S_2, where S_1, S_2 are constants such that $S_1 > S_2$. Find the mean and standard deviation of the strength of rods of length l.

(Camb. N.S.)

19. The phase θ of a source of sound of strength A is a random variable distributed uniformly in probability over the interval 0 to 2π. If the effect as heard is $x = A\cos\theta$, find the mean and variance of the distribution of x.

(Camb. N.S.)

20. If $\ln x$ is normally distributed with mean μ and standard deviation σ, prove that the mean of x is $\exp[\mu + \frac{1}{2}\sigma^2]$. (Camb. N.S.)

21. A component has a failure-time X, which is a random variable continuously distributed with $\mathscr{F}(x) = \text{prob}(X > x)$. The function $e(x)$ is defined as the expected future life of a component given to be of age x and not to have failed, i.e. $e(x) = \mathscr{E}(X - x \mid X > x)$.

Prove that

$$e(x) = \frac{1}{\mathscr{F}(x)} \int_x^\infty \mathscr{F}(y)\,dy.$$

Show that if the probability density function of X is e^{-x} $(x > 0)$, then $e(x) = 1$, whereas if the probability density function is xe^{-x} $(x > 0)$, then $e(x)$ is a decreasing function of x. (Lond. B.Sc.)

22. Construct a realization of a Poisson process in the following way:† let $\lambda = 1$ and, from tables of random sampling numbers and natural logarithms, obtain a random sample $(t_1, t_2, ..., t_n)$ from the probability density e^{-t}: let the time between the rth and $(r+1)$st incident be t_{r+1}. Divide the time scale into equal non-overlapping intervals of length 2 and compare the empirical distribution of the number of incidents in these intervals with the Poisson distribution of mean 2. Compare also the empirical distribution of the intervals between every other incident with the density te^{-t}.

23. In a Poisson process, exactly one incident is known to have occurred in a fixed interval. Show that the incident is equally likely to have occurred anywhere in the interval.

24. A Geiger counter records the arrival of α-particles (each arrival an incident). The particles arrive in a Poisson process of rate λ and if the particle finds the counter 'alive' it is recorded, a process which takes the machine time s, a random variable with density $\mu e^{-\mu s}$. During this recording time the counter is 'dead'. A particle which arrives and finds the counter 'dead' is completely ignored by it. Find the expected interval between successive *recorded* arrivals. What proportion of particles will not be recorded by the process?

25. In the presence of a steady stream of radiation of α-particles the probability that just n particles will hit the sensitive part of a Geiger counter in a time interval of length t is

$$e^{-\lambda t}(\lambda t)^n/n!,$$

where λ is a constant.

After a particle has hit *and been recorded by* the counter, the counter has a dead period of length ϵ during which no impacts are recorded and any impact does not extend the dead time. Show that the probability that fewer than n particles will be recorded in an interval of length T is

$$\frac{\lambda^n}{(n-1)!}\int_{T-(n-1)\epsilon}^{\infty} t^{n-1}e^{-\lambda t}\,dt \quad (T \geqslant (n-1)\epsilon).$$

(Camb. N.S.)

26. Ten test-tubes of nutrient material are each inoculated with 1 c.c. of a liquid containing an average of three bacteria per c.c. Under reasonable assumptions, which should be clearly stated, find the probability that exactly seven test-tubes will show bacterial growth. (Camb. N.S.)

27. Mice are injected with micro-organisms, each mouse being given a dose consisting of equal proportions of two variants A, B. It may be assumed that the numbers of effective organisms of the two variants per dose vary independently in Poisson distributions of mean μ. The mouse

† This method is explained in detail in § 3.5.

will survive if and only if there are no effective organisms in the dose. Dead mice are examined to find whether they contain organisms of one or both variants. Prove that the probability that a dead mouse contains organisms of only one variant is

$$\frac{2}{1+e^{\mu}}.$$

[It may be assumed that effective organisms present in the dose will be detected on analysis of the dead mouse.] (Lond. B.Sc.)

28. A plant whose constitution is known to be either A or B is tested by raising n offspring from its seeds. If the constitution is A then each offspring has a chance 1/4 of being white and 3/4 of being coloured, but if the constitution is B each offspring has an equal chance 1/2 of being white or coloured. Write down the probability α of exactly r offspring being white when the plant is A, and the probability β of exactly r offspring being white when the plant is B. The plant is assigned to the class A if $\alpha > \beta$, and to the class B if $\alpha < \beta$. Show that this procedure is the same as classifying the plants on the basis of the sign of the expression $(r-kn)$ where

$$k = 1-\frac{\ln 2}{\ln 3}.$$

Show that if 48 offspring are raised a plant which is truly of the constitution A will be classified as B if it produces 18 or more white offspring. Assuming the normal approximation to the binomial distribution to hold for this case, show that the probability of misclassification of A plants is about $2\frac{1}{2}$ %. (Camb. N.S.)

29. Use the normal approximation to the binomial to obtain approximations to (the values in brackets are the exact values to 5 decimals):

(*a*) The probability of 7 *or fewer* heads in 10 tosses of a fair coin (0·94531);

(*b*) the probability of *fewer than* 45 heads in 100 tosses of a fair coin (0·13563);

(*c*) the probability of *more than* 16 plants of genotype *aa* in a breeding programme involving 52 plants when each plant has independently the same chance, 1/4, of being genotype *aa* (0·13220);

(*d*) the probability of *exactly* 16 plants in the situation described in (*c*) (0·07669).

30. For the exponential distribution with density $\lambda e^{-\lambda x}$, $x \geqslant 0$, $\lambda > 0$, compare the exact probability that a random variable with that distribution departs by more than c (> 1) standard deviations from its mean, with the bound c^{-2} given by Chebychev's inequality. Determine also the multiple of the standard deviation that needs to replace c to make the exact probability c^{-2}.

Is there a distribution for which Chebychev's inequality becomes an equality?

31. At each turn of a game consisting of n independent turns, the probability that a player scores one point is p and the probability that he scores no point is $q = 1-p$. Show that the average value of the amount by which the player's total score in a game differs from np is

$$2pq\left(\frac{\partial f}{\partial p}-\frac{\partial f}{\partial q}\right) = 2s\binom{n}{s}p^s q^{n-s+1},$$

where
$$f = \sum_{r=s}^{r=n}\binom{n}{r}p^r q^{n-r}$$

and s is the least integer greater than np. (Camb. Trip.)

32. Loaves produced at a bakery have weights normally distributed with mean μ and standard deviation σ, where μ may be set at any desired value by adjusting the machinery, but σ is fixed. All loaves lighter than a certain minimum weight, m, are rejected and the remainder sold. Find the ratio, R, of the number of loaves produced for sale from a large quantity of dough to the number that would have been produced had each loaf weighed exactly m.

If $\sigma/m = k$ show that the value of μ that maximizes R is $m+\lambda\sigma$, where λ satisfies the equation

$$\frac{1}{\sqrt{(2\pi)}}\exp\{-\tfrac{1}{2}\lambda^2[\Phi(\lambda)]^{-1}-k(1+\lambda k)^{-1}\} = 0.$$

Prove that when k is small, λ is approximately $[-\ln(2\pi k^2)]^{\frac{1}{2}}$. (Camb. N.S.)

33. In a one-dimensional random walk problem, an insect is assumed in every second to move in one direction, called 'forward', with probability p_1, or to remain at rest with probability p_2, or to move in the opposite direction with probability p_3. Movements in different seconds are supposed independent. In n seconds it moves forward r_1 times, remains at rest r_2 times and moves back r_3 times. Prove by induction or otherwise that

$$P(r_1, r_2, r_3) = \frac{n!}{r_1!\,r_2!\,r_3!}p_1^{r_1}p_2^{r_2}p_3^{r_3}.$$

The total advance in position in n seconds is denoted by r $(r = r_1-r_3)$; find the mean and variance of r.

If n is large obtain limits within which r will lie with 95% probability; if you assume normality for any quantity indicate why you expect this assumption to be justified. (Camb. N.S.)

34. Compare the normal approximation, with and without a continuity correction, with the exact probability that there are more than six correct answers to ten true–false questions, under the hypothesis that each answer has a chance of 1/2 of being correct, independently for the different questions. (Lond. B.Sc.)

35. It is desired that a certain job shall be completed at time t_0. If it is not completed until time t $(t > t_0)$, there is a loss to the producer of amount $c_1(t-t_0)$, where c_1 is a positive constant. If the job is completed before t_0, at time t ($t < t_0$), there is a loss $c_2(t_0 - t)$ arising from storage costs, where c_2 is another positive constant. Suppose that the actual time of completion is normally distributed with mean $t_0 + \Delta$ and unit variance. Find the mean loss as a function of Δ. (Lond. B.Sc.)

36. Find the probability, moment and cumulant generating functions for a variable which is $G(q)$.

37. Find the probability, moment and cumulant generating functions for a variable which is $B(n, p)$. Show that

$$\mu'_r = \left(p\frac{\partial}{\partial p}\right)^r (p+q)^n.$$

38. Prove that the moment generating function of a variable distributed as $\Gamma(n, \lambda)$ is $(1 - z/\lambda)^{-n}$ for $z < \lambda$.

39. In a sequence of n random trials with constant probability p of success let u_n be the probability of an even number of successes. Prove the recurrence formula

$$u_n = qu_{n-1} + p(1 - u_{n-1}) \quad (n = 1, 2, \ldots),$$

with u_0 taken to be 1. Deduce that

$$A(z) = \frac{1 - qz}{(1-z)[1-(q-p)z]},$$

where

$$A(z) = \sum_{s=0}^{\infty} u_s z^s,$$

and hence find an explicit formula for u_n.

40. Under a newly proposed motor insurance policy, the premium is £α in the first year. If no claim is made in the first year, the premium is £$\lambda\alpha$ in the second year where $0 < \lambda < 1$ and λ is fixed. If no claim is made in the first or second years, the premium is £$\lambda^2\alpha$ in the third year; and, in general, if no claim is made in any of the first r years ($r \geqslant 1$), the premium is £$\lambda^r\alpha$ in the $(r+1)$st year.

If in any year a claim is made, the premium in that year is unaffected but the next year's premium reverts to £α, and this year is then treated as if it were the first year of the insurance for the purpose of calculating further reductions. Assuming that the probability that no claim will arise in any year is constant and equal to q, prove that in the nth year ($n \geqslant 2$) of the policy, the probabilities that the premium paid is

$$£\lambda^{n-1}\alpha \quad \text{or} \quad £\lambda^{n-j-1}\alpha \quad (1 \leqslant j \leqslant n-1)$$

are q^{n-1} and $(1-q)\,q^{n-j-1}$ respectively.

Hence calculate the expected amount of the premium payable in the nth year and show that if this mean must always exceed $k\alpha$ ($k > 0$), then

$$\lambda > \frac{k+q-1}{kq}.$$

(Leicester.)

41. A continuous random variable x has the probability density function proportional to $e^{-x}(1+x)^2$ in the range $-1 \leqslant x < \infty$, and zero otherwise.

Derive the cumulant generating function of x and hence show that γ_1 and γ_2, the coefficients of *skewness* and *kurtosis* for the distribution, are $2/\sqrt{3}$ and 2 respectively.

Also show that $\tilde{m}$, the median of the distribution, satisfies the relation

$$e^{(\tilde{m}+1)} = 1+(\tilde{m}+2)^2.$$

$[\gamma_1 = \kappa_3/\kappa_2^{\frac{3}{2}},\ \gamma_2 = \kappa_4/\kappa_2^2.]$ (Leicester.)

42. A ceramic part for an electrical unit is manufactured to order, and as a concession to the manufacturer, the customer ordering N items is prepared to accept at the same price up to αN items, where α is a given constant > 1. However, the manufacturer must supply at least the number ordered. Because of production hazards, the number of items S of approved saleable quality is a fraction x $(0 \leqslant x \leqslant 1)$ of the initial number I put into moulds. On the basis of past experience of the production process, it can be assumed that x is approximately distributed in the form with probability density

$$f(x) = 495x^8(1-x)^2 \quad \text{for} \quad 0 \leqslant x \leqslant 1.$$

If, in general, the initial number I is equal to mN $(m > \alpha)$, determine as a function of m and α the probability that: (i) the number of saleable items S will be $< N$; and (ii) the number of saleable items will be $> \alpha N$.

Hence show that the equation for m in the special case in which the manufacturer desires his risk of not meeting the minimum demand to the full to be half his risk of exceeding the maximum number of saleable items is

$$m^{11} = 55m^2(2+\alpha^9)-99(2+\alpha^{10})m+45(2+\alpha^{11}).$$

(Leicester.)

3

PROBABILITY DISTRIBUTIONS—SEVERAL VARIABLES

The ideas of the last chapter are generalized to the case where an observation consists of two real numbers, and then to the case of more than two.

3.1. The discrete case

Associated with each elementary event a in the sample space is a pair of integers; A_{rs} is the set of a for which the associated pair is (r, s) and $p_{rs} = p(A_{rs})$. Clearly

$$p_{rs} \geqslant 0, \quad \sum_{r,s} p_{rs} = 1. \tag{1}$$

A sequence satisfying (1) is said to be a *bivariate (probability) density* and the *joint density* of the random variables $\tilde{r}$ and $\tilde{s}$. The random variables are said to have a *joint distribution.* The corresponding *joint distribution function* is

$$P_{rs} = \sum_{i \leqslant r} \sum_{j \leqslant s} p_{ij}. \tag{2}$$

As with one variable this definition may usefully be extended to define $F(x, y) = p(\tilde{r} \leqslant x, \tilde{s} \leqslant y)$ for all real x and y: clearly $F(x, y) = P_{rs}$, where r and s are the integral parts of x and y respectively. $F(x, y)$ has possible discontinuities along the lines $x = r$, $y = s$ for integer r and s.

The relations between the joint density of $\tilde{r}$ and $\tilde{s}$ and the densities of $\tilde{r}$ and of $\tilde{s}$ are easily found. Let A_r be the set of a for which the first member of the pair has the value r: then $A_r = \sum_s A_{rs}$ (cf. §1.2 for the notation), and the events on the right-hand side are exclusive. Hence the density of $\tilde{r}$ is

$$p(A_r) = \sum_s p(A_{rs}) = \sum_s p_{rs} = p_{r.}, \quad \text{say.} \tag{3}$$

(We shall often have occasion to use a 'dot' as a suffix: it replaces a suffix, here s, over which summation has taken place.) Similarly, the density of $\tilde{s}$ is

$$\sum_r p_{rs} = p_{.s}. \tag{4}$$

The distributions corresponding to (3) and (4) are often called the *marginal distributions* of the joint distribution.

Consider the conditional probability $p(A_{rs} \mid A_r)$ for *fixed* r and *all* s. If $p_r \neq 0$ it equals (theorem 1.4.2) $p(A_{rs})/p(A_r) = p_{rs}/p_r$. which we write $p(s \mid r)$. Clearly

$$p(s \mid r) \geqslant 0, \quad \sum_s p(s \mid r) = 1, \tag{5}$$

so satisfying the conditions (equation 2.1.2) for a density: it is called the density of the random variable $\tilde{s}$ when the other random variable, $\tilde{r}$, has the value r, or shortly, the *conditional density* of $\tilde{s}$ for $\tilde{r} = r$. Similarly, $p(r \mid s) = p_{rs}/p_{.s}$ is the conditional density of $\tilde{r}$ for $\tilde{s} = s$. The notation $p(s \mid r)$ is an obvious extension of earlier notation.

The random variables $\tilde{r}$ and $\tilde{s}$ are *independent* if, for all r, s, $p_{rs} = p_{r.}p_{.s}$. A necessary and sufficient condition for independence is that p_{rs} be the product of a function of r and a function of s. For if $p_{rs} = f_r g_s$, summation over s, possible since $\{p_{rs}\}$ forms a convergent series, gives $p_{r.} = f_r g$, where $g = \sum_s g_s$, and over r gives $p_{.s} = f g_s$, where $f = \sum_r f_r$. But summation over both r and s shows that $fg = 1$ whence, eliminating f_r and g_s, we have $p_{rs} = p_{r.}p_{.s}$. The converse is immediate.

The mean of the conditional density (5) is

$$\sum_s s p(s \mid r) = \mathscr{E}(\tilde{s} \mid r), \tag{6}$$

say, and is called the *conditional expectation* of $\tilde{s}$ for $\tilde{r} = r$. It is also called the *regression* of $\tilde{s}$ on $\tilde{r}$. Notice that it is a function of r. Similarly,

$$\mathscr{E}(\tilde{r} \mid s) = \sum_r r p(r \mid s).$$

The expectation of $\tilde{s}$ has already been defined in §2.1 as $\sum_s s p_{.s}$

and may be written in terms of the joint density as $\sum_{r,s} sp_{rs}$. Generally if $f(r,s)$ is any function of r and s the *expectation* of $f(\tilde{r}, \tilde{s})$ is defined as

$$\mathscr{E}[f(\tilde{r},\tilde{s})] = \sum_{r,s} f(r,s)\,p_{rs}, \tag{7}$$

provided the double series is absolutely convergent. The expectation

$$\mathscr{E}[\{\tilde{r}-\mathscr{E}(\tilde{r})\}\{\tilde{s}-\mathscr{E}(\tilde{s})\}] = \mathscr{C}(\tilde{r},\tilde{s}) \tag{8}$$

is called the *covariance* of $\tilde{r}$ and $\tilde{s}$. The ratio

$$\mathscr{C}(\tilde{r},\tilde{s})/\mathscr{D}(\tilde{r})\,\mathscr{D}(\tilde{s}) = \rho(\tilde{r},\tilde{s}) \tag{9}$$

is called the *correlation coefficient* between $\tilde{r}$ and $\tilde{s}$, or simply the *correlation of* $\tilde{r}$ and $\tilde{s}$. The symmetric matrix

$$\begin{pmatrix} \mathscr{D}^2(\tilde{r}) & \mathscr{C}(\tilde{r},\tilde{s}) \\ \mathscr{C}(\tilde{r},\tilde{s}) & \mathscr{D}^2(\tilde{s}) \end{pmatrix} \tag{10}$$

is called the *variance–covariance*, or *dispersion*, matrix of $\tilde{r}$ and $\tilde{s}$.

The *joint probability generating function* of $\tilde{r}$ and $\tilde{s}$ is defined as

$$\Pi(x,y) = \sum_{r,s} p_{rs} x^r y^s, \tag{11}$$

an obvious generalization of equation 2.6.1, using a double series and two variables, x and y. The other definitions of §2.6 similarly generalize; thus the characteristic function is obtained from (11) by putting $x = e^{it}$, $y = e^{iu}$, t and u real, giving

$$\Psi(t,u) = \sum_{r,s} p_{rs} e^{i(tr+us)} = \mathscr{E}(e^{i(t\tilde{r}+u\tilde{s})}). \tag{12}$$

Several random variables

The definitions introduced in this section are all straightforward generalizations of definitions for a single random variable given in chapter 2. In most practical situations the observation made on the experimental material is not a single number but a set of numbers. If we consider the sample space in its full complexity as containing a complete description of the experimental material (§1.2), then each possible result, each elementary event, has, as part of its description, the actual observations made: thus with each *a* are associated the numbers called random variables. For example, in an investigation into mildew on apple trees, counts were made of the number of affected shoots at bud-burst and at three later occasions during the year. Each of these counts is a random variable in the sense discussed in the last chapter. The discussion there applied to each count separately, but the point of the four counts was to assess the relationships between them; for example, was a high count later in the year associated with a high count at bud-burst? If so the infection was local, if not it probably came from a distance. The definitions of this section are designed to provide the language in which such questions can be answered.

The trinomial distribution

As an example of a discrete random variable we used the number of times an event occurs in a random sequence of n trials with constant probability of success and obtained the binomial distribution (§2.1). Consider, under the same conditions, two *exclusive* events, A and B, with probabilities p and q respectively. Let C denote the event, neither A nor B, with probability $1-p-q$. A, B and C are exclusive and exhaustive. If A and B occur r and s times in the n trials, so that C necessarily occurs $(n-r-s)$ times, an argument parallel to that in §2.1 shows that

$$p_{rs} = \frac{n!}{r!\,s!\,(n-r-s)!} p^r q^s (1-p-q)^{n-r-s} \tag{13}$$

for non-negative integers r, s with $r+s \leq n$. This is the density of the *trinomial* distribution: generally, for several exclusive events, we have the *multinomial* distribution. p_{rs} is the coefficient of $x^r y^s$ in the expansion of $(px+qy+1-p-q)^n$. It is easy to verify that $p_{r.}$ (equation (3)), is $B(n, p)$ and $p_{.s}$ (equation (4)), is $B(n, q)$.

Conditional distributions

The densities of $\tilde{r}$ and $\tilde{s}$ separately (equations (3) and (4)), can be derived from the joint density of $\tilde{r}$ and $\tilde{s}$, but the converse is not true in general. The reason for this is that the joint density expresses how $\tilde{r}$ and $\tilde{s}$ influence each other, which the separate densities cannot. This influence is most completely expressed through the conditional densities $p(r \mid s)$ and $p(s \mid r)$. In the apple mildew example just mentioned, it would be natural to consider the distribution of infection later in the year for those trees with little or no infection at bud-burst and compare it with the distribution for those trees which had been heavily infected then. If $\tilde{r}$ and $\tilde{s}$ are the counts at bud-burst and later respectively, the required distributions are of $\tilde{s}$ for different, fixed values of $\tilde{r}$. This is the conditional distribution of $\tilde{s}$ for fixed $\tilde{r}$, defined by the density $p(s|r)$. Since $p_{rs} = p_{r.}p(s|r)$ it follows that the density of $\tilde{r}$ and the conditional densities of $\tilde{s}$ for each value of $\tilde{r}$ are together equivalent to the joint density. Hence the conditional densities characterize the relationship between the two variables. The same is true with $\tilde{r}$ and $\tilde{s}$ interchanged, for equally $p_{rs} = p_{.s}p(r|s)$, but in most applications one way is more natural than another: for example, it would be unnatural to consider the distribution of infection at bud-burst for trees with high infection later in the year; earlier events influence later ones, not the other way round.

An important special case is where the conditional distribution $p(s|r)$ is the same for all r; for example, where the infection later in the year is not influenced by that at bud-burst. We may then write $p(s|r) = q_s$, say, so that $p_{rs} = p_{r.}q_s$; which, when summed over r, gives $p_{.s} = q_s$, since $\sum_r p_{r.} = 1$. Hence if $p(s|r)$ is the same for all r, $p_{rs} = p_{r.}p_{.s}$, and conversely. This is a symmetrical relation so $p(r|s)$ must be the same for all s, and we say that the random variables are independent (compare the motivation for the definition of independent events, § 1.3). An important special case of independent random variables arises when we take a random sample of size 2 from a population (§ 1.3). If $\tilde{r}$ is the measurement on the first and $\tilde{s}$ on the second member of the sample then, because of the way the samples have been taken, $\tilde{r}$ and $\tilde{s}$ will, for an infinite population, be independent (see § 5.1).

In the trinomial example we have, since $\tilde{r}$ is $B(n, p)$,

$$p(s|r) = \frac{n!}{r!\,s!\,(n-r-s)!}p^r q^s (1-p-q)^{n-r-s} \Big/ \frac{n!}{r!\,(n-r)!}p^r(1-p)^{n-r}$$

$$= \frac{(n-r)!}{s!\,(n-r-s)!}\left(\frac{q}{1-p}\right)^s \left(\frac{1-p-q}{1-p}\right)^{n-r-s}, \qquad (14)$$

which is $B(n-r, q/(1-p))$. This is otherwise obvious since with r fixed only the $n-r$ events in which A does not occur are considered and $p(B|\bar{A}) = p(B)/p(\bar{A}) = q/(1-p)$. The density certainly does depend on r so that $\tilde{r}$ and $\tilde{s}$ in the trinomial distribution are not independent.

Regression

When a joint distribution has been expressed by means of a distribution of one variable and a set of conditional distributions, the features (§2.4) of these separate distributions can be used to describe the joint behaviour. The mean of the conditional density is by far the most important (equation (6)). The term regression was introduced by Francis Galton in connexion with heights of fathers ($\tilde{r}$) and sons ($\tilde{s}$): it was empirically found in that situation that if $r > \mathscr{E}(\tilde{r})$ then $\mathscr{E}(\tilde{s}|r) < r$, or fathers of above (below) average height have on the average sons shorter (taller) than themselves, or the heights *regress* towards the common mean $\mathscr{E}(\tilde{r}) = \mathscr{E}(\tilde{s})$. It is important to notice that $\mathscr{E}(\tilde{s}|r)$ is a number which is a function of r: hence with each elementary event is associated r and hence a number $\mathscr{E}(\tilde{s}|r)$, so that the conditional expectation is a random variable. In particular we may take its expectation,†

$$\begin{aligned}\mathscr{E}[\mathscr{E}(\tilde{s}|\tilde{r})] &= \sum_r \mathscr{E}(\tilde{s}|r)p_{r.} = \sum_r [\sum_s sp(s|r)]\,p_{r.} \\ &= \sum_r \sum_s sp_{rs} = \mathscr{E}(\tilde{s}). \end{aligned} \tag{15}$$

Manipulations like this are often useful in determining an expectation in stages and are analogous to applications of the generalized addition law for probabilities (theorem 1.4.4). For example, by (15), with $\tilde{r}\tilde{s}$ replacing $\tilde{s}$,

$$\mathscr{E}(\tilde{r}\tilde{s}) = \mathscr{E}[\mathscr{E}(\tilde{r}\tilde{s}|\tilde{r})] = \mathscr{E}[\tilde{r}\mathscr{E}(\tilde{s}|\tilde{r})] \tag{16}$$

since, in the conditional expectation within square brackets, $\tilde{r}$ is a constant and equation 2.4.1 applies. If $\mathscr{E}(\tilde{s}|r) = \alpha+\beta r$ for constants α and β, the regression is said to be *linear* and β is the

† Here is an example where the distinction between $\tilde{r}$ and r is necessary. $\mathscr{E}(\tilde{s}|r)$, meaning the expectation of $\tilde{s}$ when $\tilde{r} = r$, becomes $\mathscr{E}(\tilde{s}|\tilde{r})$ in (15), the expectation of $\tilde{s}$ as a function of a, through $\tilde{r}$.

(linear) *regression coefficient* of $\tilde{s}$ on $\tilde{r}$. The variances of the conditional distributions are often used,

$$\mathscr{D}^2(\tilde{s}\,|\,r) = \mathscr{E}[\{\tilde{s}-\mathscr{E}(\tilde{s}\,|\,r)\}^2\,|\,r]. \tag{17}$$

If $\mathscr{D}^2(\tilde{s}\,|\,r)$ is constant for all r, a particularly important case, the conditional densities are said to be *homoscedastic*; otherwise they are *heteroscedastic*.

A useful generalization of linear regression is to the case where the conditional expectation of $\tilde{s}$ for fixed r is a polynomial in r of specified degree. The coefficient of r^2 is the quadratic regression coefficient, etc. This case is discussed in §8.6(c).

In the trinomial case $\mathscr{E}(\tilde{s}\,|\,r)$ follows easily, since the conditional densities are binomial, and is $(n-r)\,q/(1-p)$. The regression is linear and the regression coefficient is $-q/(1-p)$. The densities are heteroscedastic since $\mathscr{D}^2(\tilde{s}\,|\,r)$ is $(n-r)\,q(1-p-q)/(1-p)^2$, depending on r.

Correlation

The conditional densities, and in particular the regressions, usually provide the most useful way of describing the relations between two random variables but there are situations in which it is not desirable to have a description which is unsymmetric in the two variables. For example, in education it is not usually more natural to consider the regression of English marks on Arithmetic marks than of Arithmetic on English. In such circumstances a feature which is symmetric in the variables and describes their joint behaviour would be wanted. The covariance satisfies these requirements. If $\tilde{r}$ and $\tilde{s}$ are independent the covariance (equation (8)) is zero, for then $p_{rs} = p_{r\cdot}\,p_{\cdot s}$ and

$$\mathscr{C}(\tilde{r},\tilde{s}) = \sum_r \{r-\mathscr{E}(\tilde{r})\}p_{r\cdot}\,.\sum_s \{s-\mathscr{E}(\tilde{s})\}p_{\cdot s} = 0 \tag{18}$$

since each sum separately is zero. On the other hand, if $\tilde{r}$ and $\tilde{s}$ are positively associated the covariance is positive, for when $\tilde{r}$ is greater (smaller) than its expectation, $\tilde{s}$ will tend to be greater (smaller) than its expectation, and the product in square brackets in (8) is positive. Conversely if, as $\tilde{r}$ increases, $\tilde{s}$ decreases, then the product is negative. The correlation coefficient is introduced because it is unaffected by a change of units in either variable: it will later be proved (§3.3) that $|\rho(\tilde{r}, \tilde{s})| \leqslant 1$. It is usually true (and always true for the bivariate normal distribution (§3.2)) that a correlation coefficient is larger in modulus the more closely the two variables are associated: in particular, by (18), it is zero when the two variables are independent, but the converse is not true. To illustrate this let $\tilde{r}$ have a distribution with zero mean and zero third moment ($\mathscr{E}(\tilde{r}^3) = 0$), and let $\tilde{s} = \tilde{r}^2$. Then a little calculation shows that $\rho(\tilde{r}, \tilde{s}) = 0$, but $\tilde{r}$ and $\tilde{s}$ are very far from being independent. If $\rho(\tilde{r}, \tilde{s}) = 0$ the variables are said to be *uncorrelated.*

The dispersion matrix, together with the expectations, $\mathscr{E}(\tilde{r})$ and $\mathscr{E}(\tilde{s})$, often provides a fair description of the joint distribution of the two variables. The advantages of writing the variances and covariances in matrix form will appear later (§3.5). The covariance is most easily calculated from the result

$$\mathscr{C}(\tilde{r}, \tilde{s}) = \mathscr{E}(\tilde{r}\tilde{s}) - \mathscr{E}(\tilde{r})\,\mathscr{E}(\tilde{s}), \tag{19}$$

which is proved by the same methods as was theorem 2.4.1. It follows that if $\tilde{r}$ and $\tilde{s}$ are independent $\mathscr{E}(\tilde{r}\tilde{s}) = \mathscr{E}(\tilde{r})\mathscr{E}(\tilde{s})$, by (18). The relationship between the covariance and a regression follows easily from (16) and (19):

$$\mathscr{C}(\tilde{r}, \tilde{s}) = \mathscr{E}[\tilde{r}\{\mathscr{E}(\tilde{s}\,|\,\tilde{r}) - \mathscr{E}(\tilde{s})\}]. \tag{20}$$

In particular if the regression is linear, $\mathscr{E}(\tilde{s}\,|\,r) = \alpha + \beta r$; then $\mathscr{E}(\tilde{s}) = \alpha + \beta\mathscr{E}(\tilde{r})$ by (15) and

$$\mathscr{C}(\tilde{r}, \tilde{s}) = \mathscr{E}[\alpha\tilde{r} + \beta\tilde{r}^2 - \alpha\tilde{r} - \beta\tilde{r}\mathscr{E}(\tilde{r})] = \beta\mathscr{D}^2(\tilde{r}). \tag{21}$$

The joint generating functions are mainly used in the important case where the random variables are independent, for then

$$\Pi(x, y) = \sum_{r,s} p_{r.}p_{.s}x^r y^s = \Pi_r(x)\,\Pi_s(y), \tag{22}$$

the product of the two generating functions of $\tilde{r}$ and $\tilde{s}$. Conversely, if $\Pi(x, y)$ factorizes, the factors are proportional to the two generating functions and the variables are independent: for if

$$\Pi(x, y) = f(x)\,g(y),$$

and we put $x = 1$, the left-hand side is $\Pi_s(y)$ and the right-hand side is a constant multiple of $g(y)$, so $\Pi_s(y) \propto g(y)$ and similarly $\Pi_r(x) \propto f(x)$. With $x = y = 1$ the product of the constants must be 1 and (22) follows. (Compare the necessary and sufficient condition for independence given above.)

The covariance of $\tilde{r}$ and $\tilde{s}$ in the trinomial distribution can be found, directly, using (19), or, since the regression is linear, by substituting the values of the regression coefficient $\beta = -q/(1-p)$ (above) and the binomial variance of $\tilde{r}$, $np(1-p)$, into (21). It gives $\mathscr{C}(\tilde{r}, \tilde{s}) = -npq$. The negative value is to be expected since the more times event A occurs the fewer opportunities there are for B to occur. The correlation is $-[pq/(1-p)(1-q)]^{\frac{1}{2}}$. The probability generating function is $(px+qy+1-p-q)^n$.

Poisson process

The definitions of this section are particularly useful when applied to stochastic processes and we illustrate with the simple example of the Poisson process (§2.3). Let $\tilde{n}(t)$ denote the number of incidents in $(0, t)$: we refer to the stochastic process $\{\tilde{n}(t), t \geqslant 0\}$. In §2.3 the Poisson distribution of $\tilde{n}(t)$, for fixed t, was obtained: consider now the joint distribution of $\tilde{n}(t)$ and $\tilde{n}(t+\tau)$, $\tau \geqslant 0$. The fundamental result is that

$$\tilde{n}(t+\tau) = \tilde{n}(t) + \tilde{m}(\tau),$$

where $\tilde{m}(\tau)$, the number of incidents in $(t, t+\tau)$, and $\tilde{n}(t)$ are independent, and $\tilde{m}(\tau)$ has the same distribution as $\tilde{n}(\tau)$, namely $P(\lambda\tau)$: this follows from the definition of the Poisson process. Consequently the conditional distribution of $\tilde{n}(t+\tau)$ for fixed $\tilde{n}(t)$ has density, for $s \geqslant r$,

$$\begin{aligned} p[\tilde{n}(t+\tau) = s \,|\, \tilde{n}(t) = r] &= p[\tilde{m}(\tau) = s-r] \\ &= e^{-\lambda\tau}(\lambda\tau)^{s-r}/(s-r)!, \end{aligned} \tag{23}$$

and hence the joint density is

$$\begin{aligned} p[\tilde{n}(t+\tau) = s, \tilde{n}(t) = r] \\ = e^{-\lambda(t+\tau)}\lambda^s t^r \tau^{s-r}/r!(s-r)!. \end{aligned} \tag{24}$$

Notice that the joint density is naturally expressed through the conditional densities (23) expressing the position 'now', at $t+\tau$, in terms of a fixed value in the 'past', at t. Clearly

$$\mathscr{E}[\tilde{n}(t+\tau)\,|\,\tilde{n}(t)] = \tilde{n}(t) + \mathscr{E}[\tilde{m}(\tau)] = \tilde{n}(t) + \lambda\tau,$$

so that the regression is linear, with regression coefficient of unity, and, by (21), $\mathscr{C}[\tilde{n}(t), \tilde{n}(t+\tau)] = \mathscr{D}^2[\tilde{n}(t)] = \lambda t$ and the correlation is therefore $\{t/(t+\tau)\}^{\frac{1}{2}}$. This correlation decreases with τ from 1 when $\tau = 0$ (and $\tilde{n}(t+\tau) = \tilde{n}(t)$) to 0 as $\tau \to \infty$, expressing the fact that the further apart in time the two counts $\tilde{n}(t)$ and $\tilde{n}(t+\tau)$ are, the less is the correlation between them. For *any* stochastic process $\tilde{n}(t)$, the correlation so obtained is called the *autocorrelation* function, a function of t and τ. The other conditional distribution, of $\tilde{n}(t)$ on $\tilde{n}(t+\tau)$, is interesting: from (24) and the $P[\lambda(t+\tau)]$ distribution of $\tilde{n}(t+\tau)$ it follows that

$$p[\tilde{n}(t) = r\,|\,\tilde{n}(t+\tau) = s] = \binom{s}{r}\left(\frac{t}{t+\tau}\right)^r\left(\frac{\tau}{t+\tau}\right)^{s-r}, \tag{25}$$

which is $B[s, t/(t+\tau)]$. We leave the reader to think out for himself why this is otherwise obvious from the definition of the Poisson process.

Generating functions

The following results will be required in §4.1.

Theorem 1. *If* $\Pi(y\,|\,r)$ *is the probability generating function of* $\tilde{s}$ *for fixed* $\tilde{r} = r$ *and* $\Pi(y)$ *is the probability generating function of* $\tilde{s}$, *then*

$$\Pi(y) = \sum_r \Pi(y\,|\,r)p_{r.}.$$

We use the same argument as led to (15)

$$\begin{aligned}\Pi(y) = \mathscr{E}(e^{\tilde{s}v}) &= \mathscr{E}.\mathscr{E}(e^{\tilde{s}v}\,|\,\tilde{r}) \\ &= \mathscr{E}.\Pi(y\,|\,\tilde{r}) \\ &= \sum_r \Pi(y\,|\,r)p_{r.}.\end{aligned}$$

Theorem 2. *The joint probability generating function,* $\Pi(x, y)$, *of* $\tilde{r}$ *and* $\tilde{s}$ *is given by*

$$\Pi(x, y) = \sum_r \Pi(y\,|\,r)p_{r.}x^r.$$

For
$$\Pi(x, y) = \sum_{r,s} p_{rs} x^r y^s = \sum_{r,s} p(s\,|\,r) p_{r.} x^r y^s$$
$$= \sum_r p_{r.} x^r \sum_s p(s\,|\,r) y^s$$
as required.

3.2. The continuous case

Associated with each elementary event a in the sample space is a pair of real numbers. This association defines a pair of random variables, $\tilde{x}$, $\tilde{y}$. Let $B_{x,y}$ be the set of a for which both $\tilde{x} \leqslant x$ and $\tilde{y} \leqslant y$, where x and y are any real numbers. Write $p(B_{x,y}) = F(x, y)$. Then the random variables $\tilde{x}$ and $\tilde{y}$ are said to have a *joint distribution* with $F(x, y)$ as their *joint distribution function.* Suppose that there exists a function $f(x, y)$ such that

$$F(x, y) = \int_{-\infty}^{x} du \int_{-\infty}^{y} dv \,.\, f(u, v), \tag{1}$$

then $f(x, y)$ is called the *joint density* of the random variables. Clearly

$$f(x, y) \geqslant 0 \quad \text{and} \quad \int_{-\infty}^{\infty} dx \int_{-\infty}^{\infty} dy \,.\, f(x, y) = 1, \tag{2}$$

and any function satisfying (2) is called a *joint* (or *bivariate*) *density.* It follows from the fundamental theorem of the integral calculus that if $f(x, y)$ is continuous then

$$f(x, y) = \partial^2 F(x, y)/\partial x\, \partial y. \tag{3}$$

Let A be any set of a defined in terms of $\tilde{x}$ and $\tilde{y}$ alone. Another theorem of the integral calculus gives

$$p(A) = \iint_A dx\, dy \,.\, f(x, y). \tag{4}$$

In particular if A is the set with $\tilde{x} \leqslant x$, (4) or (1) gives

$$p(\tilde{x} \leqslant x) = F_1(x) = \int_{-\infty}^{x} du \int_{-\infty}^{\infty} dv \,.\, f(u, v), \tag{5}$$

where $F_1(x)$ is the distribution function of $\tilde{x}$. Comparing (5) with equation 2.2.3 we see that the *marginal* density

$$f_1(x) = \int_{-\infty}^{\infty} f(x, y)\, dy \tag{6}$$

is the density of $\tilde{x}$: similarly

$$f_2(y) = \int_{-\infty}^{\infty} f(x, y)\, dx$$

is the density of $\tilde{y}$ (cf. equation 3.1.4). By analogy with the discrete case, if $f_1(x) \neq 0$, the ratio

$$f(x, y)/f_1(x) = f_2(y \mid x), \quad \text{say,}$$

considered as a function of y for fixed x, is called the *conditional density* of $\tilde{y}$ for $\tilde{x} = x$. It obviously satisfies the conditions

$$f_2(y \mid x) \geqslant 0, \quad \int_{-\infty}^{\infty} f_2(y \mid x)\, dy = 1$$

for a function to be a density. The immediate result that

$$f_2(y) = \int_{-\infty}^{\infty} f_2(y \mid x) f_1(x)\, dx \tag{7}$$

is an extension of the generalized addition law (theorem 1.4.4) with the summation replaced by an integration and the probabilities by densities. Interchanging the roles of x and y we also define $f_1(x \mid y) = f(x, y)/f_2(y)$ as the conditional density of $\tilde{x}$ for $\tilde{y} = y$. Bayes's theorem (1.4.6) and its corollary have similar generalizations,

$$f_2(y \mid x) = f_1(x \mid y) f_2(y) \Big/ \int_{-\infty}^{\infty} f_1(x \mid y) f_2(y)\, dy \tag{8}$$

or

$$f_2(y \mid x) \propto f_1(x \mid y) f_2(y), \tag{9}$$

the constant of proportionality, $f_1(x)^{-1}$, only involving x.

The random variables are *independent* if

$$f(x, y) = f_1(x) f_2(y). \tag{10}$$

The definitions of the discrete case in §3.1, from equation 3.1.6 onwards, extend to the continuous case in the obvious way, replacing summations by integrations.

Distribution functions and densities

The definitions of this section extend those of the previous section in the same way that those of §2.2 extended those of §2.1. In the continuous case we have to start with the distribution function and obtain the density from it. The fundamental property of a density in the one-dimensional case is that when integrated over a region (this is nearly always an interval) it gives the probability that the random variable lies in that region (interval). Since there are two variables involved here the integrals must be double integrals and the fundamental property remains true with the substitution of double for single integrals (equation (4)). In the bivariate case the regions are often more general than intervals and (4) is a significant extension of (1). For example, if $\tilde{x}$ and $\tilde{y}$ are the horizontal and vertical distances of the fall of an arrow on a vertical target from the centre, we may be interested in A, the set for which $\tilde{x}^2+\tilde{y}^2 \leqslant c^2$, where c is the radius of the 'bull'; $p(A)$ is then the probability of scoring a 'bull'. If the region of integration is a small interval ($x < \tilde{x} \leqslant x+h, y < \tilde{y} \leqslant y+k$) then the probability of lying in the interval is approximately $f(x, y)hk$, that is, the density times the area (replacing length) of the interval. It is helpful to think of $f(x, y) = z$ as the third axis in a set of orthogonal axes, with x and y as the other two. If z is vertical $f(x, y)$ may be thought of as a 'mountain' above the 'plane' $z = 0$. The total 'volume' of the mountain above any region in the (x, y)-plane is the probability of lying in that region. Alternatively one can think of a metal plate in the (x, y)-plane with mass density $f(x, y)$ the point (x, y): (4) is then the mass of the part A of the plate. The discrete case may be similarly regarded, with 'mass' p_{rs} at the point in the (x, y)-plane with integer co-ordinates r and s.

The distribution function is not much used in the bivariate situation but its basic properties may be noted. In the discrete case $F(x, y)$ has possible discontinuities along $x = r$ and $y = s$, where r and s are integers. Except on these lines it is constant. In the continuous case, however, it has no discontinuities since it can be written as an integral. These two cases do not exhaust the possibilities. It can happen that one variable $\tilde{x}$, say, is discrete and the other, $\tilde{y}$, is continuous. Then $F(x, y)$ is discontinuous at $x = r$ for some integral r. The distribution is then best defined by the discrete density $\{p_r\}$ for the first variable, and the conditional (continuous) density for the second, the latter only being defined for $x = r$, an integer: or alternatively by a (continuous) density $f_2(y)$ and a (discrete) conditional density for $\tilde{x}$, $p_r(y)$, say. We met an example of this situation in proving theorem 2.3.3 where we used the continuous random variable $\tilde{s}$, the time to the first incident in $(0, t)$, and the discrete random variable, there $\tilde{n}$, the number of incidents in $(0, t)$. The joint distribution was defined in the second way and equation 2.3.5 corresponds to (7) above.

Another possibility is that both variables are measured continuously and the distribution function is continuous but does not satisfy (1): this was mentioned as a possibility in the one variable case (§2.2), where it is a mathematical curiosity, but in the bivariate case it can occur in practice. That is, we show that there exists a continuous $F(x, y)$ *not* satisfying (1). Suppose $\tilde{x}$ is a continuous random variable with distribution function $F_1(x)$, and $\tilde{y} = \tilde{x}$. Then $F(x, y) = F_1(z)$ where z is the smaller of x and y. Suppose (1) to be possible. Fixing y in (1) and allowing x to vary in any interval not containing y we see that since $F(x, y)$ is constant, $f(x, y)$ must be zero. Hence $f(x, y) = 0$ everywhere, except possibly on $x = y$. But whatever value $f(x, y)$ takes on that line cannot affect the integral, hence the integral must vanish, contradicting (1). This anomaly need not worry us in this book (in practice it will only occur if the two variables are functionally related).

We leave the reader to verify that any joint distribution function has the properties:

$$\left.\begin{aligned} &F(x+h, y+k) - F(x, y+k) - F(x+h, y) + F(x, y) \geqslant 0, \\ &\text{for } h, k \geqslant 0, \text{ and} \\ &\lim_{x\to-\infty} F(x, y) = \lim_{y\to-\infty} F(x, y) = 0, \quad \lim_{x, y\to+\infty} F(x, y) = 1. \end{aligned}\right\} \quad (11)$$

Any function satisfying (11) is called a joint (or bivariate) distribution function.

Conditional distributions

The conditional densities are defined by analogy with the discrete case: the idea being that they should have similar properties to the discrete probabilities $p(s \mid r)$, etc. Their practical use is exactly the same as in the discrete case, namely to compare the distributions of one variable for different values of the other. A direct definition in terms of conditional probabilities is not possible because the equivalent of $p_{r.}$, the probability that the random variable takes a given value, is zero and hence it cannot be used as the denominator in $p(s|r) = p_{rs}/p_{r.}$. So the conditional density is *defined* as $f(x, y)/f_1(x) = f_2(y|x)$ and shown to have properties analogous to $p(s|r)$, of which (7) is the basic one. The extension of Bayes's theorem in (8) and (9) enables the arguments of § 1.6 to be extended to cases where the hypotheses form a continuous family and so are not enumerable. This will be dealt with in chapters 5–8 but we can already indicate an example of the extension. In discussing the Poisson process the probabilities usually contain the parameter λ, the rate of occurrence of incidents, and any probability is strictly conditional on a fixed value of λ. λ is part of the constant conditioning event, usually omitted. For example, $p_n(t)$, theorem 2.3.3, is, in full, the probability of n incidents in $(0, t)$, given a Poisson process of parameter λ. Given the Poisson process and t as understood we could rewrite this as $p(n|\lambda)$. This is a conditional (discrete) density for n, given λ. Now λ is continuous and may have a distribution derived from degrees of belief. In which case it becomes a random variable and has a density $p(\lambda)$, say. Then (9) says that

$$p(\lambda|n) \propto p(n|\lambda)\, p(\lambda) \tag{12}$$

and expresses the beliefs about λ after a count of n incidents has been made in time t. Equation (12) should be compared with equation 1.6.1. This problem will be discussed in detail in § 7.3.

The definition of independence has the same motivation as before: if $f_1(x|y)$ does not depend on y then it must equal $f_1(x)$ and hence $f(x, y) = f_1(x) f_2(y)$. This final symmetric form is taken as the definition corresponding to the idea that the conditional distribution of $\tilde{x}$ is not affected by $\tilde{y}$.

Expectation

The definition of expectation of a function of $\tilde{x}$ and $\tilde{y}$, $g(\tilde{x}, \tilde{y})$ say, is obviously

$$\int dx \int dy \,.\, g(x, y)\, f(x, y) = \mathscr{E}[g(\tilde{x}, \tilde{y})],$$

and of conditional expectation

$$\int g(y) f_2(y \mid x)\, dy = \mathscr{E}[g(\tilde{y}) \mid \tilde{x} = x].$$

Their properties and practical interpretations are as in the discrete case. The analogy between a bivariate distribution and the distribution of mass in the form of a plate for the continuous case and point masses in the discrete one, sheds some light on the variances and covariances. The variances will be the moments of inertia about the co-ordinate axes if the centre of gravity is taken at the origin (equivalent to taking the means as zero). The covariance will be the mixed moment of inertia with respect to these two axes.

Generating functions

As in the univariate case (§2.6) the probability generating function is not used with continuous random variables but the other generating functions are defined in the obvious way. For example, the joint characteristic function is

$$\Psi(t, u) = \int dx \int dy \,.\, e^{i(tx+uy)} f(x, y) = \mathscr{E}[e^{i(t\tilde{x}+u\tilde{y})}]$$

and the power-series expansion of its logarithm generates the cumulants.

The bivariate normal distribution

We define a special joint density of considerable importance by letting both the conditional and unconditional densities be normal. Suppose that $\tilde{x}$ is $N(\mu_1, \sigma_1^2)$, normal with expectation μ_1 and variance σ_1^2. Suppose that the conditional densities of $\tilde{y}$ for fixed $\tilde{x}$ are also normal. To define the joint distribution we have only to specify the means and variances of these conditional densities. The simplest assumptions are that the regression is *linear*

$$\mathscr{E}(\tilde{y} \mid x) = \alpha + \beta x \tag{13}$$

and *homoscedastic*,

$$\mathscr{D}^2(\tilde{y} \mid x) = \sigma^2, \tag{14}$$

where α, β and σ^2 are constants. The joint density is then the product of the two normal densities; the marginal, $N(\mu_1, \sigma_1^2)$, and the conditional, $N(\alpha+\beta x, \sigma^2)$. Hence

$$f(x, y) = (2\pi\sigma\sigma_1)^{-1}\exp\left[-\frac{1}{2}\left\{\frac{(x-\mu_1)^2}{\sigma_1^2}+\frac{(y-\alpha-\beta x)^2}{\sigma^2}\right\}\right]. \quad (15)$$

This may be written in a more symmetrical form by putting

$$\mu_2 = \alpha+\beta\mu_1, \quad (16)$$

and using this to replace α in (15). After some algebra we obtain for $f(x, y)$ the result

$$\frac{1}{2\pi\sigma_1\sigma_2\sqrt{(1-\rho^2)}}\exp\left[-\frac{1}{2(1-\rho^2)}\right.$$
$$\left.\times\left\{\frac{(x-\mu_1)^2}{\sigma_1^2}-\frac{2\rho(x-\mu_1)(y-\mu_2)}{\sigma_1\sigma_2}+\frac{(y-\mu_2)^2}{\sigma_2^2}\right\}\right], \quad (17)$$

where $\sigma_2^2 = \sigma^2+\beta^2\sigma_1^2$ and $\rho = \beta\sigma_1/\sigma_2$. (18)

The reason for the two substitutions, (18), will appear below. The density (17) is known as the *bivariate normal density*. It is of great importance in education and physical anthropology, to mention only two uses. Two measurements on adult males, such as height and weight are found to have such a density: examination marks in different subjects can often be scaled so that the density is normal. It has been derived here in what seems to be the most natural way of generalizing the single variable form.

Since (17) is unaltered by simultaneously interchanging x and y, σ_1 and σ_2, μ_1 and μ_2, and $\tilde{x}$ is $N(\mu_1, \sigma_1^2)$; $\tilde{y}$ must be $N(\mu_2, \sigma_2^2)$. Similarly, since $\tilde{y}$ for $\tilde{x} = x$ is $N(\alpha+\beta x, \sigma^2)$, that is

$$N[\mu_2+(x-\mu_1)\,\rho\sigma_2/\sigma_1,\ \sigma_2^2(1-\rho^2)], \quad (19)$$

the distribution of $\tilde{x}$ for $\tilde{y} = y$ must be

$$N[\mu_1+(y-\mu_2)\,\rho\sigma_1/\sigma_2,\ \sigma_1^2(1-\rho^2)], \quad (20)$$

so that the regression of $\tilde{x}$ on $\tilde{y}$ is also linear and homoscedastic. The regression coefficients are $\beta = \rho\sigma_2/\sigma_1$ and $\rho\sigma_1/\sigma_2$ respectively. The covariance is $\beta\sigma_1^2$, by equation 3.1.21, which equals $\rho\sigma_1\sigma_2$ by (18). Hence the correlation coefficient (equation 3.1.9) is ρ (hence the notation). The bivariate normal distribution is therefore defined by five *parameters*, the means, μ_1 and μ_2, and the dispersion matrix (equation 3.1.10),

$$\begin{pmatrix} \sigma_1^2 & \rho\sigma_1\sigma_2 \\ \rho\sigma_1\sigma_2 & \sigma_2^2 \end{pmatrix}. \tag{21}$$

The quadratic in braces in (17) is a general quadratic and the five parameters are those needed to define it, bearing in mind that the integral being one imposes a constraint, in particular that it be positive semi-definite. For this distribution ρ is a reasonable measure of the association between $\tilde{x}$ and $\tilde{y}$ and is extensively used in education.

It is often convenient to introduce another random variable $\tilde{\varepsilon}_y$ defined by

$$\tilde{y} = \alpha + \beta\tilde{x} + \tilde{\varepsilon}_y. \tag{22}$$

Now, for $\tilde{x} = x$, $\tilde{y}$ is $N(\alpha+\beta x, \sigma^2)$, hence, still for fixed x, $\tilde{y}-\alpha-\beta x$ is $N(0, \sigma^2)$ (§2.5). In other words the distribution of $\tilde{y}-\alpha-\beta x = \tilde{\varepsilon}_y$ is the same for all x, or $\tilde{x}$ and $\tilde{\varepsilon}_y$ are independent. Hence the total variation of $\tilde{y}$ is ascribable to two *independent* causes: first, the association with $\tilde{x}$ (the $\alpha+\beta\tilde{x}$ term) and secondly, $\tilde{\varepsilon}_y$. If the variation is measured by means of the variance, we have

$$\mathscr{D}^2(\tilde{y}) = \sigma_2^2 = \beta^2\sigma_1^2 + \sigma^2, \tag{23}$$

equation (18). σ^2 is called the *residual variance* of $\tilde{y}$, allowing for $\tilde{x}$: it is the variation in $\tilde{y}$ that is not ascribable to $\tilde{y}$'s association with $\tilde{x}$. From (18), $\sigma^2 = \sigma_2^2(1-\rho^2)$, so the residual variance of $\tilde{y}$ is a proportion $(1-\rho^2)$ of the total variance of $\tilde{y}$. (The same holds true for $\tilde{x}$.) Thus if $\rho = 0$ none is ascribable to the other, and if $\rho = \pm 1$ all of it is and the joint density ceases to exist. In fact if $\rho = 0$ the density, (17), factorizes and hence $\tilde{x}$ and $\tilde{y}$ are independent, and conversely. Thus with *normal* variables, the terms 'uncorrelated' and 'independent' are synonymous. The breakdown of the variation into two components is an important idea and is much used in the technique of analysis of variance (§ 6.5).

The joint moment generating function of $\tilde{x}$ and $\tilde{y}$ is the obvious generalization of equation 2.6.10, namely

$$\Phi(z, w) = \iint dx\,dy \,.\, f(x, y)\, e^{zx+wy},$$

with $f(x, y)$ given by (15). We know the means of $\tilde{x}$ and $\tilde{y}$ are μ_1 and μ_2 respectively, so it is more useful to generate the moments about the means by considering $e^{-\mu_1 z - \mu_2 w}\,\Phi(z, w)$ (see §2.6). It is simple, but tedious, to evaluate the double integral by completing the squares in $(x-\mu_1)$ and $(y-\mu_2)$, along the lines used in the univariate case (equation 2.6.13), to obtain

$$\exp\,[-\mu_1 z - \mu_2 w]\,\Phi(z, w) = \exp\,[\tfrac{1}{2}(z^2\sigma_1^2 + 2zw\rho\sigma_1\sigma_2 + w^2\sigma_2^2)].$$

The cumulant generating function, $\ln\Phi(z, w)$, is thus

$$\mu_1 z + \mu_2 w + \tfrac{1}{2}\sigma_1^2 z^2 + \rho\sigma_1\sigma_2 zw + \tfrac{1}{2}\sigma_2^2 w^2$$

and $\kappa_{20} = \sigma_1^2$, $\kappa_{02} = \sigma_2^2$, $\kappa_{11} = \rho\sigma_1\sigma_2$ (in an obvious notation) and all higher cumulants vanish.

These results may more easily be obtained by using conditional generating functions as in theorem 3.1.1. We have

$$\Phi(z, w) = \int e^{zx} f_1(x) \left[\int e^{wy} f_2(y\,|\,x)\,dy\right] dx.$$

Since $f_2(y\,|\,x)$ is normal the integral in square brackets has already been evaluated in the univariate normal case (equation 2.6.13) and we have

$$\exp\,[-z\mu_1 - w\mu_2]\,\Phi(z, w)$$
$$= \exp\,[\tfrac{1}{2}w^2\sigma^2] \int \exp\,[(z + w\beta)(x - \mu_1)]\,f_1(x)\,dx.$$

This integral is again that for a univariate normal generating function, so using equation 2.6.13 we have

$$\exp\,[-z\mu_1 - w\mu_2]\,\Phi(z, w) = \exp\,[\tfrac{1}{2}(z + w\beta)^2\sigma_1^2 + \tfrac{1}{2}w^2\sigma^2],$$

which gives the same result on substituting for β and σ^2 using (18).

More than two variables

All the ideas of these last two sections generalize without any difficulty and without any essentially new concepts, at least as far as is needed in this book, to the case of any finite number of variables. The only point that requires attention is the definition of independence. The random variables $(\tilde{x}_1, \tilde{x}_2, \ldots, \tilde{x}_n)$ are *independent* if their joint density $f(x_1, x_2, ..., x_n)$ is the product of their separate densities $f_i(x_i)$, that is

$$f(x_1, x_2, ..., x_n) = f_1(x_1) f_2(x_2) ... f_n(x_n). \quad (24)$$

The motivation is as usual in terms of conditional densities (cf. (25)). The discrete case is similar. The definition should be compared with that for independent events (§1.3): there it was necessary to consider subcollections of events. This is not necessary with random variables since (24) implies the similar result for subcollections by integration of (24) with respect to variables not in the subcollection. An alternative definition of independence for events can be obtained by introducing random variables equal to 1 if A_n obtains and 0 if $\bar{A}_n$ obtains (the notation of §1.3): the events are independent iff the random variables are. The random variables of an infinite sequence are independent if those of every finite set are.

The general form of (24) in the dependent case is

$$f(x_1, x_2, ..., x_n) = f_1(x_1) f_2(x_2 | x_1) f_3(x_3 | x_1, x_2) ... f_n(x_n | x_1, x_2, ..., x_{n-1}), \quad (25)$$

where, for example, $f_3(x_3 | x_1, x_2)$ is the conditional density of $\tilde{x}_3$ given $\tilde{x}_1$ and $\tilde{x}_2$. Compare the generalized multiplication law, theorem 1.4.8.

The normal distribution also extends but its study is postponed until §3.5 because additional results are needed.

The definitions of these last two sections are introduced in order to be able to handle several random variables: we are now in a position to prove some important results using these ideas. From this point onwards the distinction between $\tilde{x}$ and x or $\tilde{r}$ and r—the distinction between a random variable and the values it takes—will no longer be made. Once the ideas of expectation, and particularly conditional expectation, have been understood the manipulations can most easily be carried out without having to consider whether $\tilde{x}$ or x is intended.

3.3. Linear functions of random variables

Throughout this section x_i will denote a random variable; a_i, b_i, a and b will denote constants. By a *linear function* of

$(x_1, x_2, \ldots, x_n)$ is meant a function of the form $\sum_{i=1}^{n} a_i x_i + b$: if $b = 0$ it is a *homogeneous* linear function. All summations are from 1 to n.

Theorem 1. $\mathscr{E}(\sum_i a_i x_i + b) = \sum_i a_i \mathscr{E}(x_i) + b.$

We give the proof for two variables x_1, x_2 and consider the continuous case with density $f(x_1, x_2)$. The left-hand side is, by definition, equation 3.1.7,

$$\int dx_1 \int dx_2 (a_1 x_1 + a_2 x_2 + b) f(x_1, x_2),$$

which may be written as the sum of three integrals, the first of which is

$$\int dx_1 \int dx_2 \, a_1 x_1 f(x_1, x_2) = a_1 \mathscr{E}(x_1).$$

The other two follow similarly and the result is proved for $n = 2$. The case of general n follows by induction. Discrete variables may be handled similarly, replacing integrations by summations.

Theorem 2.

$$\mathscr{D}^2(\sum_i a_i x_i + b) = \mathscr{D}^2(\sum_i a_i x_i) = \sum_i a_i^2 \mathscr{D}^2(x_i) + \sum_{i \neq j} a_i a_j \mathscr{C}(x_i, x_j).$$

Again consider $n = 2$ and the continuous case. Other cases follow as with theorem 1. The first equality is equation 2.4.4 and it remains only to prove the second. Let $x_i' = x_i - \mathscr{E}(x_i)$, then $\mathscr{D}^2(x_i') = \mathscr{E}(x_i'^2)$ and $\mathscr{C}(x_i', x_j') = \mathscr{E}(x_i' x_j')$. Also by theorem 1 $\mathscr{E}(\Sigma a_i x_i) = \Sigma a_i \mathscr{E}(x_i)$ so $\mathscr{D}^2(\Sigma a_i x_i) = \mathscr{D}^2(\sum_i a_i x_i')$. Hence

$$\begin{aligned}\mathscr{D}^2(\Sigma a_i x_i') &= \mathscr{E}[(a_1 x_1' + a_2 x_2')^2] \\ &= \mathscr{E}[a_1^2 x_1'^2 + 2a_1 a_2 x_1' x_2' + a_2^2 x_2'^2] \\ &= a_1^2 \mathscr{E}(x_1'^2) + 2a_1 a_2 \mathscr{E}(x_1' x_2') + a_2^2 \mathscr{E}(x_2'^2)\end{aligned}$$

by a further use of theorem 1. This proves the result.

Corollary. *If the* $\{x_i\}$ *are uncorrelated*

$$\mathscr{D}^2(\sum_i a_i x_i) = \sum_i a_i^2 \mathscr{D}^2(x_i).$$

If they are uncorrelated the covariances are all zero.

Theorem 3. *If the $\{x_i\}$ are uncorrelated and $\mathscr{E}(x_i) = \mu$, $\mathscr{D}^2(x_i) = \sigma^2$, for all i; then if $\bar{x} = n^{-1} \sum_i x_i$*

$$\mathscr{E}(\bar{x}) = \mu, \quad \mathscr{D}^2(\bar{x}) = \sigma^2/n.$$

The results are special cases of theorem 1 and the corollary to theorem 2 obtained by putting $a_i = n^{-1}$.

Theorem 4. *If the $\{x_i\}$ are uncorrelated and have the same variance, then the random variables $\sum_i a_i x_i$ and $\sum_i b_i x_i$ are uncorrelated iff $\sum_i a_i b_i = 0$.*

To prove them uncorrelated we have to show that the covariance is zero. Using the same ideas as in the proof of theorem 2, we see that this covariance is

$$\mathscr{C}(\sum_i a_i x'_i, \sum_i b_i x'_i) = \mathscr{E}[(\Sigma a_i x'_i)(\Sigma b_i x'_i)]$$

$$= \mathscr{E}[\sum_{ij} a_i b_j x'_i x'_j].$$

But if $i \neq j$, $\mathscr{E}(x'_i x'_j) = 0$, and $\mathscr{E}(x'^2_i) = \sigma^2$, say, is the same for all i, hence the covariance is $(\Sigma a_i b_i)\sigma^2$ which proves the result.

Combination of observations

The subject treated in this and some later sections (particularly §6.6) used to be called 'Combination of Observations'. This is a study of the way in which random variables can be combined, starting here with linear combinations, the most widely used. The first theorem is almost obvious from the way expectation has been defined, and an important thing to notice is that it is true without any conditions on the random variables beyond the fact that they have expectations. For example, in the last line of the proof of theorem 2 it has been used for the random variables x'^2_1, $x'_1 x'_2$ and x'^2_2 which are functionally related to each other. The same remark applies to theorem 2 provided that the variances exist. In this result we see how the covariance enters when calculating variances of linear functions: it is similar to the use of mixed moments of inertia in calculating the moment of inertia about a general line in the body, in terms of the mixed moment and the moments about the axes (the analogues of the variances). The corollary, however, only obtains when the variables are uncorrelated, that is when the correlation coefficient (or the covariance) between any pair is zero: in particular when the variables are independent. Under these conditions the corollary states that the variance of a linear function is a *different* linear function of the variances, namely with a_i^2 replacing a_i. Notice that this is a result in terms of variances, not in terms of standard deviations, and is a good example of the way in which variances are used in calculations at the end of which a square root is taken to obtain the more readily interpreted standard deviation (cf. §2.4).

We shall find it useful in chapter 8 and in §3.5 to use vector and matrix notation, and theorem 2 can be conveniently expressed in these terms. Let **a** denote the *column* vector with elements $a_1, a_2, \ldots, a_n$ and **a**′ its transpose, a *row* vector with the same elements. Let **C** denote the dispersion matrix of the x's (equation 3.1.10). Then

$$\mathscr{D}^2(\mathbf{a}'\mathbf{x}) = \mathbf{a}'\mathbf{C}\mathbf{a},$$

a quadratic form in the a_i.

Arithmetic means

Theorem 3, although only a special case of the two previous theorems, is a famous and important result. It says that the expectation of $\bar{x}$, usually called the arithmetic mean, or simply the mean, of the $\{x_i\}$, is the common expectation of each x_i, and that the standard deviation of the mean is the common standard deviation of each x_i divided by root n. The results are true under the conditions of uncorrelated random variables each with the same expectation and variance. A common circumstance where these conditions obtain is in a *random sample* of size n (§1.3) from a large, or infinite, population, where each member of the population has associated with it a number. A random sample of size n will yield n numbers: x_1, for the first member of the sample, and so on. Clearly these n numbers, or observations, or random variables, will all have the same distribution and be independent (approximately for a finite population, exactly for an infinite one) because of the random sampling (see also §5.1). Hence, in particular, they will be uncorrelated and have the same means and variances, and the theorem can be phrased: 'The mean of a random sample of size n has expectation equal to the common expectation and standard deviation $n^{-\frac{1}{2}}$ that of each observation.'

It also applies when n repetitions have been made of a single measurement of an unknown quantity, θ, say: for example, n determinations of the acceleration due to gravity in an observatory. If $\mathscr{E}(x_i) = \theta$ the measurements are said to be *unbiased*: if not, the difference, $\mathscr{E}(x_i) - \theta$, is called the *bias* or the *systematic error*. Presumably the determinations have equal precision (though it is easy to imagine circumstances when they will not because of some improvement in design as the experiment progresses) which will be reflected in a common variance, $\mathscr{D}^2(x_i) = \sigma^2$: σ^2 is a measure of the *random error* (cf. §5.1). The total error is a combination of these two errors (compare the argument in the last section in connexion with the bivariate normal distribution). Finally the determinations will often be independent, though again not always: chemists often ignore a measurement which deviates too much from the others and substitute another one which is in better agreement with them, so introducing a correlation. If, however, all these conditions obtain, the theorem says that the mean has the same bias as each measurement (in particular it is unbiased if they are) and the random error, measured by the variance, is reduced by a factor of $1/n$. Scientists typically use repeated measurements to reduce random error; our theorem gives us a measure of how well this is done; but systematic error is not, of course, reduced by such repetition. The standard deviation is the usual measure of random error quoted,† and the factor $n^{-\frac{1}{2}}$ shows that repetition becomes less and less worth while as n increases. For example, the error is halved by taking four measurements as against one, but sixteen measurements (twelve more) are needed to reduce it by a further factor of 2.

To illustrate and emphasize the fact that theorem 3 only obtains for uncorrelated variables, consider an example with

† A term which used to be used, but is not now in favour, is *probable error*; this is one-half the interquartile range (§2.4).

$n = 2$ where all the conditions of that theorem are satisfied except that the correlation between x_1 and x_2 is ρ. Then theorem 1 is still available to show that $\mathscr{E}(\bar{x}) = \mu$ but we must use theorem 2 in order to obtain the variance: in fact $\mathscr{D}^2(\bar{x}) = \frac{1}{4}\{\mathscr{D}^2(x_1)+\mathscr{D}^2(x_2)+2\mathscr{C}(x_1, x_2)\}$, with $a_1 = a_2 = \frac{1}{2}$. Consequently $\mathscr{D}^2(\bar{x}) = \frac{1}{2}(1+\rho)\sigma^2$. With $\rho = 0$ the result of theorem 3 obtains: with $\rho = 1$ the measurements are positively perfectly correlated and they are only as good as a single measurement: with $\rho = -1$ they are negatively perfectly correlated and the value of μ is determined with no random error at all. This final result is explained by the fact that x_1 must be as far above μ as x_2 is below it in order that ρ be -1; that is $x_1 = \mu+\epsilon$, $x_2 = \mu-\epsilon$, so that $\bar{x} = \mu$ exactly. (Compare the comments at the end of this section.) If measurements can be obtained with a large negative correlation then the mean may be a very precise determination.

Experimental design

One of the tasks of a statistician is to measure the random phenomena that are the subject of his study, and one measure of the randomness is the random error just defined. Theorem 3 shows how the random error is reduced by repeated determinations of an unknown quantity. But in addition to this, the statistician has to *design* an experiment to reduce the random error and to make as precise a determination as is possible. The following famous example illustrates this design aspect and the use of the theorems of this section. Four similar objects A_1, A_2, A_3, A_4 are to be weighed on an ordinary two-pan balance provided with a set of known weights. The usual method of determining the weights of the four objects would be to weigh each one separately, using, therefore, four weighings. We show that with four weighings each object can be weighed more accurately than this. Suppose that a weighing consists of putting the object or objects of unknown weight on one or both of the pans and adding known weights to the amount x_i, say, to one of the pans until they balance. Then x_i is a random variable. The sample space is the space of all possible weighings of the object or objects and x_i is defined for each. Suppose $\mathscr{E}(x_i)$ is the true weight of the unknown(s) and $\mathscr{D}^2(x_i) = \sigma^2$. In the former assumption we are supposing that there is no bias† and in the latter that the precision does not depend on the unknown weight. If the separate weighings of each object are made and supposed independent, each unknown weight will be determined with no systematic error and with standard deviation σ. Consider instead the following series of weighings:

	Left-hand pan	Right-hand pan
First weighing	A_1, A_2	A_3, A_4
Second weighing	A_1, A_3	A_2, A_4
Third weighing	A_1, A_4	A_2, A_3
Fourth weighing	A_1, A_2, A_3, A_4	None

Suppose that in the ith weighing weights to the amount x_i have to be added to the right-hand pan to obtain balance. (In the first three weighings the weights may have to be added to the left-hand pan to obtain balance, but this can be thought of as an addition to the right-hand pan with x_i negative.) The x_i are then independent random variables with variances σ^2. If no error were present ($\sigma = 0$) the weights θ_i of A_i could be found from the equations

$$\left.\begin{aligned} \theta_1+\theta_2 &= \theta_3+\theta_4+x_1,\\ \theta_1+\theta_3 &= \theta_2+\theta_4+x_2,\\ \theta_1+\theta_4 &= \theta_2+\theta_3+x_3,\\ \theta_1+\theta_2+\theta_3+\theta_4 &= x_4, \end{aligned}\right\} \tag{1}$$

with solutions

$$\left.\begin{aligned} \theta_1 &= \tfrac{1}{4}(x_1+x_2+x_3+x_4),\\ \theta_2 &= \tfrac{1}{4}(x_1-x_2-x_3+x_4),\\ \theta_3 &= \tfrac{1}{4}(-x_1+x_2-x_3+x_4),\\ \theta_4 &= \tfrac{1}{4}(-x_1-x_2+x_3+x_4). \end{aligned}\right\} \tag{2}$$

Now suppose $\sigma \neq 0$ and the right-hand sides of these equations are used to determine the weights of the objects. (Why this should be done will be clarified in §6.6.) Then theorems 1 and 2 show that each determination is made with no systematic error and with standard deviation $\frac{1}{2}\sigma$. Hence by weighing the objects

† If there is bias, or systematic error, it may sometimes be removed by making a second weighing with the known and unknown weights interchanged in the pans. Then $\mathscr{E}(x_1) = x+b$, $\mathscr{E}(x_2) = x-b$ and so $\mathscr{E}(\bar{x}) = x$, in an obvious notation.

together in these four weighings instead of separately in four weighings the standard deviation has been halved. Another way of halving the standard deviation would be to weigh each object separately four times, sixteen weighings in all. The design proposed here uses only a quarter of this effort and is said to be four times as efficient as that using four separate weighings.

This little experiment also illustrates another idea of some practical consequence. If there are several related things to be investigated it is usually better, if possible, to combine them in one experiment rather than to perform separate experiments. Here it is better to use all four objects in each experiment (weighing) than to use them singly. Another application of the same idea arises when it is required to determine the values of certain variables, e.g. temperature, pressure, humidity, etc., which maximize the yield of a product: it is better to vary the variables together in a systematic way, rather than to vary one factor at a time.

Finally, in connexion with this experiment, notice that theorem 4, which applies to uncorrelated variables of equal variance, such as the x_i in (2), shows that the θ_i in the same equations are uncorrelated. If it was possible to suppose the θ_i normally distributed—and, as will be seen later (§3.6) this might be very reasonable—then they would be independent. Thus this experiment retains the feature of the 'one-at-a-time' method of making independent determinations. This is a desirable feature since if the determinations were, for example, positively correlated then one over-determination might mean that they were all over-determined. The condition for lack of correlation in theorem 4 is the same as the condition for the vectors $(a_1, a_2, \ldots, a_n)$, $(b_1, b_2, \ldots, b_n)$ to be orthogonal. We shall find it useful to think of the vectors associated with linear forms (§8.3); by analogy with these, the determinations of (2) are said to be *orthogonal* and the experiment is called an *orthogonal experiment*. This example will be discussed again in §6.6.

Correlation coefficient

It is now possible to prove that the correlation coefficient (equation 3.1.9) never exceeds one in modulus. Let x and y be two random variables with correlation ρ and $\mathscr{D}^2(x) > 0$, and consider the variance of $ax+y$, which must be non-negative. By theorem 2,

$$\mathscr{D}^2(ax+y) = a^2\mathscr{D}^2(x)+2a\mathscr{C}(x, y)+\mathscr{D}^2(y).$$

Considered as a function of a this is a quadratic which tends to $+\infty$ as $|a| \to \infty$ and must have no distinct real roots, since if it did the variance would be negative for values of a between the roots. The condition for this is

$$\mathscr{C}^2(x, y) \leqslant \mathscr{D}^2(x)\,\mathscr{D}^2(y),$$

which is equivalent to $|\rho| \leqslant 1$. $\rho = \pm 1$ iff there is an a_0 with $\mathscr{D}^2(a_0 x+y) = 0$, so that $a_0 x+y = c$, where c is a constant: that is, x and y are linearly related.

The results of this section for means and variances extend without difficulty of principle to the higher moments, but the results are not so often required.

3.4. Approximate means and variances

Theorem 1. *If x is a random variable with $\mathscr{E}(x) = \mu$, $\mathscr{D}^2(x) = \sigma^2$, and $y = \phi(x)$ then, for sufficiently small σ, and well-behaved ϕ*

$$\mathscr{E}(y) \doteq \phi(\mu)+\tfrac{1}{2}\sigma^2\phi''(\mu), \tag{1}$$

and

$$\mathscr{D}^2(y) \doteq \phi'(\mu)^2\,\sigma^2. \tag{2}$$

Expand $\phi(x)$ in a Taylor series about μ as far as the term† in $(x-\mu)^2$:

$$y = \phi(x) = \phi(\mu)+(x-\mu)\,\phi'(\mu)+\tfrac{1}{2}(x-\mu)^2\,\phi''(\mu) + O[(x-\mu)^3]. \tag{3}$$

Now if σ is small, by Chebychev's inequality (theorem 2.4.3), x will depart only a little from μ except on rare occasions and therefore $(x-\mu)$ will typically be small. The final term will therefore be neglected. The result (1) then follows on taking expectations of both sides of (3) using theorem 3.3.1 on the right-hand side and $\mathscr{E}(x-\mu) = 0$, $\mathscr{E}[(x-\mu)^2] = \sigma^2$.

† We write $O[g(z)]$, read 'order of $g(z)$', for a function $f(z)$ which is such that $|f(z)| < Kg(z)$ for some constant K and all z sufficiently near to some value, here zero.

To prove (2), rewrite (3) as

$$y-\phi(\mu)-\tfrac{1}{2}\sigma^2\phi''(\mu)$$
$$= (x-\mu)\,\phi'(\mu)+\tfrac{1}{2}[(x-\mu)^2-\sigma^2]\,\phi''(\mu)+O[(x-\mu)^3],$$

and square both sides, retaining only the terms of order $(x-\mu)^2$ on the right-hand side. On taking expectations the right-hand side is $\phi'(\mu)^2\sigma^2$, to the approximation used, and the left-hand side is, by (1), $\mathscr{D}^2(y)$, to the same order.

Theorem 2. *If x and y are random variables with $\mathscr{E}(x) = \mu$, $\mathscr{E}(y) = \nu$, $\mathscr{D}^2(x) = \sigma^2$, $\mathscr{D}^2(y) = \tau^2$ and $\rho(x,y) = \rho$, and $z = \phi(x,y)$ then, for sufficiently small σ and τ, and well-behaved ϕ*

$$\mathscr{E}(z) \simeq \phi(\mu,\nu)+\tfrac{1}{2}\sigma^2\frac{\partial^2\phi}{\partial x^2}+\rho\sigma\tau\frac{\partial^2\phi}{\partial x\,\partial y}+\tfrac{1}{2}\tau^2\frac{\partial^2\phi}{\partial y^2}, \tag{4}$$

and
$$\mathscr{D}^2(z) \simeq \sigma^2\left(\frac{\partial\phi}{\partial x}\right)^2+2\rho\sigma\tau\left(\frac{\partial\phi}{\partial x}\frac{\partial\phi}{\partial y}\right)+\tau^2\left(\frac{\partial\phi}{\partial y}\right)^2, \tag{5}$$

where all the partial differentials are evaluated at $x = \mu$, $y = \nu$.

The theorem is a direct generalization to two variables of the single variable case of theorem 1, and it is proved in exactly the same way using, in lieu of (3), the two-variable form of Taylor's theorem,

$$\phi(x,y) \simeq \phi(\mu,\nu)+(x-\mu)\frac{\partial\phi}{\partial x}+(y-\nu)\frac{\partial\phi}{\partial y}$$
$$+\tfrac{1}{2}(x-\mu)^2\frac{\partial^2\phi}{\partial x^2}+(x-\mu)(y-\nu)\frac{\partial^2\phi}{\partial x\,\partial y}$$
$$+\tfrac{1}{2}(y-\nu)^2\frac{\partial^2\phi}{\partial y^2}, \tag{6}$$

where the higher-order terms have been omitted and the partial differentials are evaluated at $x = \mu$, $y = \nu$.

Comments on the results

The results of this section, like those of the last, belong to the field of 'Combination of observations' and were well known long before the modern theory of probability and statistics was developed. They differ from the results of the last section in that they are only approximate, instead of exact, but on the other hand they apply to a wide class of functions besides linear ones.

The approximation is useful for relatively small standard deviations and the results are mainly of value when the measurements are fairly precise. For example, the surveyor uses them, and so does the physicist; in the biological sciences where the variation is larger more refined analysis is often needed. Theorem 1 is the single-variable form and we begin by making some remarks about the proof. The Taylor series could be extended to more terms and then the higher moments would enter into (1) and (2), but these are seldom useful: indeed even the second term in (1), $\frac{1}{2}\sigma^2\phi''(\mu)$, can often be ignored, as indeed it is effectively in the proof of (2). Notice that with that term retained the expectation of ϕ is not quite ϕ evaluated at the expectation of x, but the term represents a correction, which is usually small. In the proof we have omitted to state any conditions on $\phi(x)$, but obviously some are needed; first, to make the Taylor series expansion about $x = \mu$ possible, which requires conditions on ϕ in the neighbourhood of $x = \mu$; and secondly, to avoid anomalies of behaviour away from μ which, although of very small probability, can be of importance. As an example of the latter consider the function $\phi(x) = x^{-1}$; this could cause trouble near the origin even if the probability of values near the origin was small since when they occur $\phi(x)$ is large. We have not attempted to state such conditions but the person using the results should be aware of the possible pitfalls in the uncritical use of them. No new points arise in the extension to two variables in theorem 2 and the extension to any number is immediate. We might, however, look at another 'proof' which has at least mnemonic value and differs only in details from that given: it employs the notion of differentials, δx, small increments in x. We have

$$\delta\phi = \frac{\partial\phi}{\partial x}\,\delta x + \frac{\partial\phi}{\partial y}\,\delta y.$$

If both sides are squared and expectations taken, treating all the differentials as deviations from means, (5) immediately follows.

Example 1. The logarithmic transformation. Let x be a positive random variable and $y = \ln x$. Then from (1),

$$\mathscr{E}(y) \simeq \ln\mu - \tfrac{1}{2}\sigma^2/\mu^2, \tag{7}$$

and

$$\mathscr{D}^2(y) \simeq \sigma^2/\mu^2. \tag{8}$$

These are more conveniently expressed in terms of

$$\lambda = \sigma/\mu = \mathscr{V}(x),$$

the coefficient of variation of x: then

$$\mathscr{E}(y) \simeq \ln\mu - \tfrac{1}{2}\lambda^2, \quad \mathscr{D}^2(y) \simeq \lambda^2. \tag{9}$$

As has already been mentioned (§2.4) $\mathscr{V}(x)$ is often constant over a series of experiments; in this case the logarithm will have constant variance which is convenient for many purposes (compare, for example, §§6.1, 6.4). The result is also useful in certain applications involving formulae which are multiplicative because the logarithmic transformation makes them additive. For example, one method of determining the viscosity, η, of a fluid is to measure the pressure drop ΔP along a tube of length l and radius a when the fluid flows through at rate Q; the relevant formula is

$$\eta = \pi(\Delta P)a^4/8lQ.$$

If ΔP, a, l and Q are measured independently it follows from (8), on taking logarithms and using the exact formula for the variance of a linear function (corollary, theorem 3.3.2), that

$$\mathscr{V}^2(\eta) \simeq \mathscr{V}^2(\Delta P) + 16\mathscr{V}^2(a) + \mathscr{V}^2(l) + \mathscr{V}^2(Q).$$

If the precisions with which the four quantities ΔP, a, l and Q are measured are known, then this result enables the precision of the determination of viscosity to be calculated. Notice that because of the factor 16 in front of $\mathscr{V}^2(a)$ it is most important that the tube used be of uniform radius and that this radius be measured accurately: it would be foolish to measure the other quantities carefully and yet use a poor quality tube.

We saw in §2.4 that the coefficient of variation of a $\Gamma(n, \lambda)$ variable was $n^{-\frac{1}{2}}$, for all λ. It follows that if we have a set of random variables with Γ-distributions having a common value of n, then their logarithms will have approximately, for large n, a common standard deviation. This will prove useful in later applications (§7.3).

Example 2. *The square-root transformation.* Suppose that experiments are being carried out to investigate the density of blood corpuscles in the human body. Part of the experimental technique may consist of the preparation of a slide and a count of the number of corpuscles in the field of view of the microscope. By the theory of the Poisson process the count may be expected to have a Poisson distribution and, since the variance of a Poisson distribution is equal to its mean, it will follow that if some factor increases the mean number of corpuscles it will also increase the variance. Consequently the experimenter will have counts of different precisions which can make the inferences a little awkward. One way out of the difficulty might be to transform the counts to obtain new values whose accuracies are constant whatever shift of mean takes place. The problem then is: if x is $P(\mu)$ can we find $y = \phi(x)$ with a variance not dependent on μ? This can be done approximately by use of (2). If $\mathscr{D}^2(y) = c^2$ we have $c^2 \simeq \phi'(\mu)^2\mu$ since $\mathscr{D}^2(x) = \mu$, or, ignoring the approximation,

$$\frac{d\phi}{d\mu} = \frac{c}{\sqrt{\mu}},$$

whence $$\phi(\mu) = 2c\sqrt{\mu},$$

ignoring a possible constant. Taking $c = \frac{1}{2}$, for convenience, we have $\phi(x) = \sqrt{x}$ and the square-root transformation of $P(\mu)$ yields an approximately constant variance $c^2 = \frac{1}{4}$. The expectation of $\sqrt{x}$ is, by (1), $\sqrt{\mu}-\frac{1}{8}/\sqrt{\mu}$. Notice that the correction term $(\frac{1}{2}\sigma^2\phi''(\mu) = -\frac{1}{8}/\sqrt{\mu})$ becomes less important as $\mu \to \infty$. Similar remarks would apply to any higher terms that could be added to (2) and hence the approximation gets better as $\mu \to \infty$. Near $\mu = 0$ the variance can be made more nearly constant by using $\sqrt{(x+\frac{3}{8})}$, or $\sqrt{x}+\sqrt{(x+1)}$, instead of $\sqrt{x}$. An application is given in §7.3.

Example 3. The inverse-sine transformation. Suppose x is $B(n, p)$ and consider the same problem as in example 2 with fixed n and varying p, and try to find a transformation yielding a variance independent of p. If $\mathscr{D}^2(y) = c^2$ (possibly a function of n) we have $c^2 \simeq \phi'(np)^2 npq$ since $\mathscr{E}(x) = np$ and $\mathscr{D}^2(x) = npq$. By an argument similar to that above we easily obtain with $c = 1/2\sqrt{n}$, that $\phi(x) = \sin^{-1}\sqrt{(x/n)}$. This is called the inverse-sine transformation and gives approximately constant variance $c^2 = 1/4n$. (Notice that y is measured in radians, from the solution of the differential equation.) The approximation improves as $n \to \infty$: for small n the transformations

$$\sin^{-1}\sqrt{[(x+\tfrac{3}{8})/(n+\tfrac{3}{4})]}$$

or
$$\tfrac{1}{2}\{\sin^{-1}\sqrt{[x/(n+1)]}+\sin^{-1}\sqrt{[(x+1)/(n+1)]}\}$$

are better.

3.5. Exact methods

Theorem 1. *If x is a continuous random variable with density $f(x)$, and if $y = \phi(x)$ is a strictly monotonic differentiable function of x with continuous derivative, then the density of y exists and is given by*

$$g(y) = f(x)\,|dx/dy|. \tag{1}$$

Suppose ϕ is strictly *increasing*. Then the event $y \leqslant t$, for any real t, is the same as the event $x \leqslant s$, where s is the unique value such that $\phi(s) = t$. Hence these two events have the same probability and the distribution function of y is

$$p(y \leqslant t) = p(x \leqslant s) = \int_{-\infty}^{s} f(u)\,du = \int_{-\infty}^{t} f(u)\frac{du}{dv}\,dv,$$

where, in the final integral, the substitution $v = \phi(u)$ has been made. Hence the distribution function of y can be written as an integral and, since $du/dv > 0$, the integrand, the density, is as stated. The continuity of the derivative ensures that the change of variable in the integral is valid.

If ϕ is strictly decreasing we have similarly

$$p(y \leqslant t) = p(x \geqslant s) = \int_{s}^{\infty} f(u)\,du = -\int_{-\infty}^{t} f(u)\frac{du}{dv}\,dv,$$

and since $du/dv < 0$ the result follows.

The conditions on $\phi(x)$ ensure that to each value of y there is at most one x with $y = \phi(x)$. This is necessary in order to pass from $p(y \leqslant t)$ to $p(x \leqslant s)$ or $p(x \geqslant s)$.

If we next consider the joint density of two random variables, x and y, and consider two functions $w = \phi(x, y)$ and $z = \psi(x, y)$ whose joint density is required, we assume for the same reasons that the transformation is 1-1: that is, to each w, z there exists a unique pair x, y such that $w = \phi(x, y)$, $z = \psi(x, y)$. Satisfactory conditions for the subsequent change of variable are that the Jacobian of the transformation exists and that the partial differentials are continuous. We state the bivariate case: the general case of n random variables is similar. The proof is a straightforward extension of the one variable case.

Theorem 2. *If x, y are continuous random variables with joint density $f(x, y)$ and if $w = \phi(x, y)$, $z = \psi(x, y)$ is a* 1-1 *transformation from (x, y) to (w, z) whose Jacobian,*

$$J = \begin{vmatrix} \frac{\partial \phi}{\partial x} & \frac{\partial \phi}{\partial y} \\ \frac{\partial \psi}{\partial x} & \frac{\partial \psi}{\partial y} \end{vmatrix},$$

exists with all the partial differentials continuous, then the joint density of w and z exists and is given by

$$g(w, z) = f(x, y) J^{-1}. \tag{2}$$

Theorem 3. *If the joint density of two random variables x and y is $f(x, y)$ and $z = x+y$, then the probability density of z exists and is given by*

$$h(z) = \int_{-\infty}^{\infty} f(x, z-x)\, dx. \tag{3}$$

For any real t,

$$p(z \leqslant t) = p(x+y \leqslant t) = \iint_{x+y \leqslant t} dx\, dy\, f(x, y).$$

Changing the variable y in the integral to $z = x+y$ we obtain

$$p(z \leqslant t) = \int_{-\infty}^{t} dz \int_{-\infty}^{\infty} dx\, f(x, z-x).$$

The distribution function is expressible as an integral and the result follows.

Corollary. *If x and y are independent, so that $f(x, y) = f(x)g(y)$, say, then*

$$h(z) = \int_{-\infty}^{\infty} f(x) g(z-x)\, dx. \tag{4}$$

The equivalents of theorems 1 and 2 are without significant content in the discrete case but theorem 3 is useful, a summation replacing the integration. For example, in the corollary, if the densities are $\{p_i\}$ and $\{q_i\}$ for x and y respectively, then the density $\{r_i\}$ of $z = x+y$ is

$$r_i = \sum_j p_j q_{i-j}. \tag{5}$$

Theorem 4. *If x and y are independent random variables with characteristic functions $\Psi_x(t)$ and $\Psi_y(t)$ respectively, then $z = x+y$ has characteristic function given by*

$$\Psi_z(t) = \Psi_x(t)\,\Psi_y(t). \tag{6}$$

$$\begin{aligned}\Psi_z(t) &= \mathscr{E}(e^{itz}), \quad \text{(for the definition see §2.6)},\\ &= \mathscr{E}(e^{it(x+y)})\\ &= \iint dx\,dy\,e^{it(x+y)}f(x)g(y), \quad \text{in the notation of the corollary},\\ &= \int e^{itx}f(x)\,dx \int e^{ity}g(y)\,dy, \quad \text{as required.}\end{aligned}$$

The discrete case follows similarly.

The logarithm of $\Psi_x(t)$ is therefore additive when *independent* random variables are *summed*, and this is a reason for introducing the cumulant generating function (§2.6).

Functions of random variables

The results of the last two sections enable the means and variances of a function of random variables to be obtained either exactly (linear functions in §3.3) or approximately (§3.4) in terms of the means and dispersion matrices of the random variables. Often these provide sufficient knowledge of the distribution of the function, for example if the distribution is known to be normal, but there are situations in which more knowledge, for example the density, is required. Also there are certain exact results on combination of observations that are useful. It is to these that we turn in this section. The reader may like to be reminded before studying the proofs that a density is a function which when suitably integrated (or summed) gives a probability, both in the univariate and multivariate cases. It is often useful to have a geometrical picture in mind; for example, in the bivariate case, to consider the integration of the density over regions in the plane.

Theorem 1 is most easily remembered in the form

$$g(y)\,dy = f(x)\,dx,$$

using positive differentials, and many writers include the differential element when referring to a density (of a continuous random variable). For example, $(2\pi)^{-\frac{1}{2}}e^{-\frac{1}{2}x^2}$ is a density, $N(0, 1)$, considered as a function of x, but not as a function of x^2 (see below), so it would be written $(2\pi)^{-\frac{1}{2}}e^{-\frac{1}{2}x^2}\,dx$, to indicate this. The proof of the theorem (and of theorem 2) involves merely a change of variable in an integral and this change introduces the differential element or the Jacobian. The restrictions placed on the function are the same as those placed on the Jacobian in order that the usual formula for the change of variable be valid. The proof of theorem 1 shows the need for these restrictions since, in the general case, the event $y \leqslant t$ would correspond to some quite complicated event in terms of x and not to the simple event $x \leqslant s$. Even if y were increasing, but not strictly increasing, the result could fail and the density not exist: thus if $\phi(x) = 0$ for $0 \leqslant x \leqslant 1$ then $p(y = 0)$ is at least $\int_0^1 f(x)\,dx$ which may well be positive, and hence the distribution of y will have a discontinuity at the origin. Nevertheless, the case of general $\phi(x)$ can be dealt with by theorem 1: the method of doing this may best be explained by an example.

Example 1. *The square of a normal variable.* Let x be $N(0, \sigma^2)$ and consider the distribution of $y = x^2$. The function $\phi(x) = x^2$ does not obey the conditions of the theorem over the whole range of x, but it does so over the ranges $x > 0$, and $x < 0$, separately. So suppose $x > 0$ first. Then the contribution to the density of y is, remembering $dy/dx = 2x = 2y^{\frac{1}{2}}$, from (1)

$$(2\pi\sigma^2)^{-\frac{1}{2}}e^{-\frac{1}{2}y/\sigma^2}\,\tfrac{1}{2}y^{-\frac{1}{2}}. \tag{7}$$

There is a similar contribution from $x < 0$ which is the same as (7) because $|dy/dx| = 2|x| = 2y^{\frac{1}{2}}$, and the density for x is symmetric about $x = 0$. Hence the final density of y (for $y > 0$, necessarily) is twice (7). This may be put into a recognizable form by writing $\lambda = 1/2\sigma^2$. Then

$$g(y) = e^{-\lambda y}\lambda^{\frac{1}{2}}y^{-\frac{1}{2}}/\sqrt{\pi}, \tag{8}$$

and comparison with equation 2.3.7 shows that $y = x^2$ is $\Gamma(\frac{1}{2}, 1/2\sigma^2)$, a gamma variable with parameter $1/2\sigma^2$ and index† $\frac{1}{2}$. Notice that the mean of the normal distribution is zero, but that the variance is arbitrary.

Example 2. *The log-normal distribution.* Suppose that y is a positive random variable whose logarithm is normally distributed: what is its density? Let $x = \ln y$ and x be $N(\mu, \sigma^2)$. Then $dy/dx = y$, and $y = e^x$ is strictly increasing, so that (1) gives

$$g(y) = \frac{1}{\sqrt{(2\pi)}\,\sigma y}\exp\left\{-\frac{1}{2\sigma^2}(\ln y - \mu)^2\right\}.$$

The density is zero at $y = 0$, rises to a maximum at $y = e^{\mu-\sigma^2}$ (the mode) and tends to zero as y tends to infinity. The easiest way to find the moments is to find those of e^x directly from the normal density. For example (equation 2.6.13)

$$\mu_r' = \frac{1}{\sqrt{(2\pi)}\,\sigma}\int e^{rx}e^{-\frac{1}{2}(x-\mu)^2/\sigma^2}dx = \exp\{\tfrac{1}{2}r(2\mu + \sigma^2 r)\}.$$

We shall see later (theorem 3.6.1) that the normal distribution arises in considering the *sum* of a large number of independent and identically distributed random variables. It follows that the log-normal distribution can arise in dealing with the *product* of a large number of similar random variables. For example, the size of incomes might have a log-normal distribution (§2.4). If one has a sum of money out at a fixed rate of interest, then in a given period of time it is multiplied by a quantity: hence the total sum of money is the product of a number of variables. (It might be even more relevant to the distribution of capital.) Similarly, the size of particles in a powder often has a log-normal distribution. Let X be the size of a particle which is being ground. The grinding will produce a particle of size $\xi_1 X$, say; a second grinding will produce from the latter particle one of size $\xi_2(\xi_1 X)$, and so on. If the assumption is made that the grinding proportions ξ_i do not depend on the size of the particle or the index i, then the final size is the product of independent and identically distributed random variables.

† It was shown at the end of §2.5 that $(-\frac{1}{2})! = \sqrt{\pi}$.

Expectations

Theorem 1 also shows that two definitions of expectation given previously are consistent. In §2.2 the expectation of x was defined as $\int xf(x)\,dx$, and that of $\phi(x)$ as $\int \phi(x)f(x)\,dx$. But the first definition says that the expectation of $y = \phi(x)$ is $\int yg(y)\,dy$ and hence we have two definitions of $\mathscr{E}(y)$, one in terms of the density of y, the other in terms of the density of x. If $\phi(x)$ satisfies the conditions of the theorem then this shows that the two definitions are equivalent. The general case follows by dividing the range of x up into parts within which $\phi(x)$ does satisfy the conditions, as in example 1. The second form for the expectation is particularly useful since it does not require the evaluation of the density of ϕ. An instance of its use is the passage from the first to the second line in the proof of theorem 4. Theorem 3 could have been used to write down the expectation in terms of $h(z)$, but this definition avoids using that theorem and is more direct. Similar remarks apply in the discrete case, but then there is no differential element to complicate the picture.

Construction of random samples

An immediate corollary of theorem 1 has a special use. If $\phi(x)$ is put equal to $F(x)$, the distribution function of the random variable x, and if $F(x)$ obeys the conditions of the theorem, y—which necessarily lies between 0 and 1—has in that interval density $g(y) = 1$. This follows since $dy/dx = f(x)$. The distribution with density equal to $(b-a)^{-1}$ in the interval (a, b) with $a < b$, and otherwise zero, is called the *uniform distribution* in (a, b). In words the corollary says that the distribution function of a random variable has a uniform distribution in the interval (0, 1)—subject to the conditions on the distribution function just stated. The special use of this result is to obtain random samples from any distribution. Tables are available of random sampling numbers; for example, Kendall and Babington Smith (1954). These give sequences of the digits 0 to 9 arranged in random order, that is arranged so that the digits occur equally often, and after any sequence, say 97231, of any length, any digit is as likely to occur as any other. (For detailed discussion of the practical difficulties here, refer to the tables.) If these digits are taken in groups of n and a decimal point put in front of the group, the numbers so formed are approximately, a random sample from a uniform distribution in (0, 1)—exactly, they are grouped into intervals of width 10^{-n} (§2.4). Consequently, treating one of the numbers as y of the theorem, the values† $F^{-1}(y)$ form a random sample from the distribution function $F(x)$. Normally a table of $F(x)$ can be used backwards (i.e. from $F(x)$ read x) in order to obtain $F^{-1}(y)$. Random samples from the standardized normal distribution $\Phi(x)$ have been produced in this way by Wold (1954). We use the device in connexion with the exponential distribution in §4.1. If $F(x)$ does not obey the conditions for application of the corollary, minor modifications to the method enable it to continue to be used.

Sums of random variables

The corollary to theorem 3 is more important than the theorem itself. If $f(x)$ and $g(x)$ are any two functions the integral in (4) is called the *convolution* of $f(x)$ and $g(x)$ and is extensively studied in many branches of mathematics. We illustrate its application to the Poisson distribution, using, of course, the discrete form (5). If x is $P(\mu)$, y is $P(\nu)$ and they are independent, then the density of $x+y$ is, by (5),

$$\sum_{j=0}^{i} e^{-\mu}\mu^j e^{-\nu}\nu^{i-j}/(i-j)!\,j!.$$

The range of summation is only from 0 to i since if $x+y = i$ then neither x nor y can exceed i. This may be rewritten

$$\{e^{-(\mu+\nu)}/i!\}\sum_{j=0}^{i}\binom{i}{j}\mu^j\nu^{i-j},$$

and the sum is equal to $(\mu+\nu)^i$ by the binomial theorem. Hence the density of $x+y$ is Poisson with mean $\mu+\nu$. Or the sum of two independent Poisson variables is Poisson. This result is

† F^{-1} denotes the function inverse to F. That is, $F^{-1}[F(x)] = x$.

obvious from the results on a Poisson process (§2.3). Let x be the number of incidents in $(0, t)$ and y the number in $(t, t+u)$, with $\lambda t = \mu$, $\lambda u = \nu$, where λ is the parameter of the Poisson process. Then x and y are Poisson variables with means μ and ν (theorem 2.3.3) and, by the definition of the process, are independent. But $x+y$ is the number of incidents in $(0, t+u)$ and is therefore, by the same theorem, $P(\mu+\nu)$. The reader may like to prove directly by (4) and also by using known properties of a Poisson process that the sum of two independent Γ-variables with the same parameter has also a Γ-distribution with that parameter and index equal to the sum of the indices.

The main tools for handling the convolution of two functions are the Laplace and Fourier transforms, so in probability we use the generating function or the characteristic function. Theorem 4 shows the reason for the success of this tool: with summation of *independent* random variables the characteristic functions combine much more easily, namely by multiplication, than do the density functions in (4). Hence in dealing with sums of independent random variables the characteristic function is an important and almost indispensable tool. We illustrate with the normal distribution. The characteristic function of an $N(\mu, \sigma^2)$ variable is (equation 2.6.13 with $z = it$)

$$\exp\left[i\mu t - \tfrac{1}{2}\sigma^2 t^2\right].$$

Hence if x and y are independent and respectively $N(\mu_1, \sigma_1^2)$ and $N(\mu_2, \sigma_2^2)$ it follows from (6) that the characteristic function of $x+y$ is the product of the separate characteristic functions, namely

$$\exp\left[i(\mu_1+\mu_2)t - \tfrac{1}{2}(\sigma_1^2+\sigma_2^2)t^2\right].$$

But this is immediately recognizable as the characteristic function of an $N(\mu_1+\mu_2, \sigma_1^2+\sigma_2^2)$ variable. Since, as we explained in §2.6, the characteristic function truly characterizes the distribution (there are not two distributions with the same characteristic function) the sum of two independent normal variables is also normal. Notice how simple this proof is and compare it with the tedious algebra that the use of (4) would involve. Notice that the mean and variance of $x+y$ agree with our previous results in §3.3. The result extends to the sum of any number of independent normal variables, and also their mean, which is merely a multiple, n^{-1}, of the sum and is therefore also normal (§2.5).

It does no harm to emphasize again that theorem 4 applies only to independent variables. However, the example of the normal distribution does not require this restriction: since we need it later we state the result.

Theorem 5. *If x and y have a joint normal density then $x+y$ is also normal.*

We use the notation of §3.2. Consider the joint density of $x+y$ and, say, x. This can be found by theorem 2 from the joint density of x and y with $w = x+y$ and $z = x$. The Jacobian is unity and it is clear that the general quadratic in x and y in braces in the normal density (equation 3.2.17) remains quadratic in w and z. Hence w and z have a joint normal density and the marginal distribution of w is normal. The parameters of the normal density of $x+y$ are most easily found from theorems 3.3.1 and 3.3.2 to be

$$\mathscr{E}(x+y) = \mu_1+\mu_2, \quad \mathscr{D}^2(x+y) = \sigma_1^2+\sigma_2^2+2\rho\sigma_1\sigma_2:$$

alternatively they may be derived by obtaining the quadratic just mentioned in w and z and comparing with equation 3.2.17 and the known mean and variance of x in that distribution. Since if x is normal so is ax, for a constant a, it follows that any linear function $ax+by$ is normal if x and y jointly are. The result and its proof extend to any number of multivariate normal variables.

Multivariate normal distribution

In chapter 8 we shall require the general multivariate normal density for n variables and we introduce and discuss it at this point to illustrate methods of handling several random variables. Consider random variables $x_1, x_2, \ldots, x_n$ with densities defined as follows:

$$\left.\begin{aligned} &x_1 \text{ is } N(0, \sigma_1^2), \\ &x_i \text{ is } N\left(\sum_{j=1}^{i-1} \beta_{ij} x_j, \sigma_i^2\right), \\ &\qquad\qquad \text{for fixed } x_1, x_2, \ldots, x_{i-1} \quad (1 < i \leqslant n). \end{aligned}\right\} \tag{9}$$

In words, each random variable is normally distributed for fixed values of the previous variables, about a mean which is a homogeneous linear function of the previous variables, with constant variance. By equation 3.2.25 the joint density of the x's is

$$\left(\frac{1}{2\pi}\right)^{\frac{1}{2}n}\frac{1}{\sigma_1\sigma_2\ldots\sigma_n}\exp\left[-\frac{1}{2}\left\{\frac{x_1^2}{\sigma_1^2}+\frac{(x_2-\beta_{21}x_1)^2}{\sigma_2^2}+\ldots+\frac{(x_n-\beta_{n1}x_1-\beta_{n2}x_2-\ldots-\beta_{n,n-1}x_{n-1})^2}{\sigma_n^2}\right\}\right]. \quad (10)$$

β_{ij} is only used when $j < i$: define $\beta_{ii} = -1$ and $\beta_{ij} = 0$ for $j > i$. Also let $\gamma_{ij} = \beta_{ij}/\sigma_i$ for all i, j. Then (10) can be rewritten

$$\left(\frac{1}{2\pi}\right)^{\frac{1}{2}n}\frac{1}{\sigma_1\sigma_2\ldots\sigma_n}\exp\left[-\frac{1}{2}\sum_{i=1}^{n}\left(\sum_{j=1}^{n}\gamma_{ij}x_j\right)^2\right]. \quad (11)$$

Change to random variables

$$z_i = -\sum_j \gamma_{ij}x_j = -\sum_j \beta_{ij}x_j/\sigma_i. \quad (12)$$

The Jacobian of the transformation (see theorem 3.5.2) is the modulus of the determinant of the matrix $\mathbf{\Gamma}$ with elements γ_{ij}. But $\gamma_{ij} = 0$ for $j > i$ ($\mathbf{\Gamma}$ is said to be *lower triangular*) so that this determinant is the product of the diagonal elements, namely $\prod_i \gamma_{ii} = (-)^n \Pi\sigma_i^{-1}$. Hence the joint density of the z_i is simply

$$(2\pi)^{-\frac{1}{2}n}\exp\left[-\tfrac{1}{2}\Sigma z_i^2\right].$$

The joint density factorizes into terms containing only one of the z's, and, by the argument of §3.1, the z's are independent and clearly $N(0, 1)$ variables. Thus (12) gives a set of independent standardized variables which are, of course, easy to handle. z_i is simply the deviation of x_i from its regression $\sum_{j=1}^{i-1}\beta_{ij}x_j$, suitably standardized.

From (12) the x_i can be written as linear functions of the z_j:

$$x_i = -\Sigma\gamma^{ij}z_j,$$

where $\{\gamma^{ij}\}$ are the elements of $\mathbf{\Gamma}^{-1}$. Hence, by theorem 3.3.2,

$$\mathscr{D}^2(x_i) = \sum_j (\gamma^{ij})^2,$$

and the total variation of x_i has been expressed as the sum of a number of separate variations due to each z. This generalizes equation 3.2.23.

We may rewrite (11) as

$$(2\pi)^{-\frac{1}{2}n}(-)^n |\mathbf{\Gamma}| \exp\left[-\frac{1}{2}\sum_{i,j=1}^{n} x_i a_{ij} x_j\right], \tag{13}$$

where $a_{ij} = \sum_k \gamma_{ki}\gamma_{kj}$ and $|\mathbf{\Gamma}|$ is the determinant of $\mathbf{\Gamma}$. If $\mathbf{A}$ is the (symmetric) matrix with elements a_{ij}, then

$$\mathbf{A} = \mathbf{\Gamma}'\mathbf{\Gamma}, \tag{14}$$

where a prime denotes transpose. We now obtain a simple interpretation for $\mathbf{A}$.

From (9)

$$\mathscr{E}(x_1) = 0 \quad \text{and} \quad \mathscr{E}(x_i \mid x_1, x_2, \ldots, x_{i-1}) = \sum_{j=1}^{i-1} \beta_{ij} x_j.$$

It follows by successive use of this last result that

$$\mathscr{E}(x_i) = \mathscr{E}.\mathscr{E}(x_i \mid x_1, x_2, \ldots, x_{i-1}) = \sum_{j=1}^{i-1} \beta_{ij}\mathscr{E}(x_j) = 0$$

(using equation 3.1.15 and theorem 3.3.1). Hence from (12) and theorem 3.3.1 again,

$$\mathscr{C}(z_i, z_j) = \mathscr{E}(z_i z_j) = \mathscr{E}\left(\sum_{l=1}^{n} \gamma_{il} x_l \sum_{m=1}^{n} \gamma_{jm} x_m\right)$$

$$= \sum_{l,m} \gamma_{il}\gamma_{jm} c_{lm},$$

where $c_{lm} = \mathscr{E}(x_l x_m) = \mathscr{C}(x_l, x_m)$. Since the z_i are independent $N(0, 1)$ this result may be written in matrix notation as

$$\mathbf{I} = \mathbf{\Gamma}\mathbf{C}\mathbf{\Gamma}',$$

where $\mathbf{I}$ is the unit matrix and $\mathbf{C}$ is the matrix of elements c_{lm}. But $\mathbf{\Gamma}$ is non-singular (we found its determinant above as a product of inverses of standard deviations), so that

$$\mathbf{C} = \mathbf{\Gamma}^{-1}(\mathbf{\Gamma}')^{-1} = (\mathbf{\Gamma}'\mathbf{\Gamma})^{-1} = \mathbf{A}^{-1}. \tag{15}$$

Hence $\mathbf{A}$ is the inverse of the dispersion matrix of the x's.

Also $|\mathbf{C}|^{\frac{1}{2}} = \pm|\mathbf{\Gamma}|^{-1}$, from (15) and hence, taking the positive sign for the square root, (13) may be written

$$(2\pi)^{-\frac{1}{2}n}\,|\mathbf{C}|^{-\frac{1}{2}}\exp\left[-\frac{1}{2}\sum_{i,j=1}^{n} x_i c^{ij} x_j\right], \tag{16}$$

where c^{ij} are the elements of $\mathbf{C}^{-1} = \mathbf{A}$. We shall later (§8.3) find it useful to pass from (16), in terms of the dispersion matrix, to the regression form (10).

If x_i is replaced by $x_i - \mu_i$ then we have the general *multivariate normal density*

$$(2\pi)^{-\frac{1}{2}n}\,|\mathbf{C}|^{-\frac{1}{2}}\exp\left[-\frac{1}{2}\sum_{i,j=1}^{n}(x_i-\mu_i)c^{ij}(x_j-\mu_j)\right] \tag{17}$$

with dispersion matrix, $\mathbf{C}$, as before and $\mathscr{E}(x_i) = \mu_i$. The regressions corresponding to (9) will be

$$\mathscr{E}[(x_i-\mu_i)\,|\,x_1, x_2, \ldots, x_{i-1}] = \sum_{j=1}^{i-1}\beta_{ij}(x_j-\mu_j),$$

that is $$\mathscr{E}(x_i\,|\,x_1, x_2, \ldots, x_{i-1}) = \alpha_i + \sum_{j=1}^{i-1}\beta_{ij}x_j,$$

where $$\alpha_i = \mu_i - \sum_{j=1}^{i-1}\beta_{ij}\mu_j.$$

3.6. Limit theorems

Theorem 1. *(Central Limit Theorem.) If $\{x_n\}$ is a sequence of independent random variables each having the same distribution of mean μ and variance σ^2, then the sequence of distribution functions of $\left(\sum_{i=1}^{n} x_i - n\mu\right)\Big/(\sigma\sqrt{n})$ converges to the distribution function of an $N(0, 1)$ variable; that is to $\Phi(x)$.*

The proof is outlined below.

A sequence $\{y_n\}$ of random variables is said to *converge in probability* (or *converge weakly*) to a constant c if, given any $\epsilon > 0$,

$$\lim_{n\to\infty} p(|y_n - c| > \epsilon) = 0. \tag{1}$$

Theorem 2. *(Weak law of large numbers.) If $\{x_n\}$ is as in theorem 1, then $\bar{x} = n^{-1}\sum_{i=1}^{n} x_i$ converges weakly to μ.*

By theorem 3.3.3 $\mathscr{E}(\bar{x}) = \mu$, $\mathscr{D}^2(\bar{x}) = \sigma^2/n$, and hence by the corollary to theorem 2.4.3 (Chebychev's inequality), for any $c > 0$

$$p(|\bar{x}-\mu| > c\sigma/\surd n) \leqslant c^{-2}.$$

Rewriting this with $\epsilon = c\sigma/\surd n$

$$p(|\bar{x}-\mu| > \epsilon) \leqslant \sigma^2/\epsilon^2 n.$$

Since the right-hand side tends to zero with n, so does the left.

A sequence $\{y_n\}$ of random variables is said to *converge strongly* to a constant c if

$$p(\lim_{n\to\infty} y_n = c) = 1. \tag{2}$$

Theorem 3. *(Strong law of large numbers.) If $\{x_n\}$ is a sequence of independent random variables with a common distribution then a necessary and sufficient condition for $\bar{x}$ to converge strongly to a value μ is that $\mathscr{E}(x_n)$ exists and is equal to μ.*

This result is discussed below.

Limits of distribution functions

It is often useful to be able to approximate to one distribution by another which is more convenient to handle. We have already met one example in §2.5 where the binomial distribution was approximated to by the normal distribution. The method used by a mathematician to discuss approximations is to consider a sequence of expressions which tend to a limiting value: each expression is an approximation to the limit (or vice versa), the later ones in the sequence being typically better than the earlier ones. A familiar example is an expansion of a function in an infinite series: any finite number of terms of the series provide an approximation to the function. This was the method used in de Moivre's theorem just referred to: here the limit is an approximation to the members of the sequence. In order to use such ideas in probability we have to say what is meant by a sequence of distributions converging to a limiting distribution. In this book we shall only have occasion to consider the case of a univariate random variable and the definition is most easily and usefully expressed in terms of the distribution function which is a probability, rather than in terms of the density which is not.

Let $F_n(t)$ be a sequence of distribution functions and $F(t)$ another distribution function which is a continuous function of t: then the obvious definition of convergence of distributions is to demand that $F_n(t) \to F(t)$ as $n \to \infty$ for all t, and this is the definition always used. If $F(t)$ is not continuous it rather surprisingly turns out that this definition is unsatisfactory, but in this book we need only the continuous $F(t)$ so the difficulty is ignored. This definition has already been used in §2.5, without a formal explanation, for the case where $F(t) = \Phi(t)$, the standardized normal distribution function (which is continuous).

Now one way of establishing limit theorems for distributions is by direct calculation of the distribution functions; but this is usually tedious and a much more powerful method is to use characteristic functions. The following result is of considerable use.

Theorem 4. *A necessary and sufficient condition that a sequence of distribution functions* $\{F_n(t)\}$ *with corresponding characteristic functions* $\{\phi_n(t)\}$ *tend to a limit* $F(t)$, *for all* t, *is that* $\phi_n(t)$ *tends to a limit* $\phi(t)$ *which is continuous at* $t = 0$. $\phi(t)$ *is then the characteristic function corresponding to* $F(t)$.

The theorem is true for any $F(t)$ but only continuous $F(t)$ will be used here. The theorem replaces convergence of distribution functions by convergence of characteristic functions. Now if a sequence of distribution functions tends to a limit, it is easy to see whether the limit is a distribution function or not (equations 2.2.1 and 2). But, in general, it is not easy to see whether a function is a characteristic function. Consequently, if we could prove that a sequence of characteristic functions tended to a limit we should not know whether the limit was a characteristic function. The condition of the theorem that the limit be continuous at the origin is easily verified and ensures not only that the limit is a characteristic function but that it is the characteristic function corresponding to the limiting distribution function. In many situations the limiting behaviour of the characteristic functions is easier to establish than that of the distribution functions directly, and the result is very powerful. The proof is beyond the level of this book and will have to be left, just as the proof that the characteristic function truly characterizes the distribution has had to be omitted. An interested reader with enough mathematics can find proofs in Cramér (1946) or Loève (1960).

Central Limit Theorem

The main use of theorem 4 is to prove the classical and important central limit theorem, theorem 1. The theorem is stated in terms of distribution functions; the proof uses characteristic functions. Furthermore, as we are dealing with sums of independent random variables we can use theorem 3.5.4 which is more conveniently expressed in logarithms of characteristic functions, or cumulant generating functions. Now since x_r has finite variance σ^2 it follows from equation 2.6.9 that its cumulant generating function (with $z = it$, to ensure that it exists) is

$$\ln \mathscr{E}\{\exp[itx_r]\} = \mu it - \tfrac{1}{2}\sigma^2 t^2 + R(t),$$

where
$$R(t)/t^2 \to 0 \quad \text{as} \quad t \to 0. \tag{3}$$

Hence
$$\ln \mathscr{E}\{\exp[it(x_r - \mu)]\} = -\tfrac{1}{2}\sigma^2 t^2 + R(t),$$

on subtracting μit from each side, and

$$\ln \mathscr{E}\{\exp[it(x_r - \mu)/\sigma\sqrt{n}]\} = -\tfrac{1}{2}t^2/n + R(t/\sigma\sqrt{n})$$

on writing $t/\sigma\sqrt{n}$ in place of t. The left-hand side is the cumulant generating function of $(x_r - \mu)/\sigma\sqrt{n}$ and hence by theorem 3.5.4 the cumulant generating function of $\sum_{r=1}^{n}(x_r - \mu)/\sigma\sqrt{n}$ is the sum of the separate functions, that is

$$n[-\tfrac{1}{2}t^2/n + R(t/\sigma\sqrt{n})]. \tag{4}$$

Now because of (3), $nR(t/\sigma\sqrt{n}) \to 0$ as $n \to \infty$, so (4) tends to $-\tfrac{1}{2}t^2$ which is certainly continuous at the origin and so is a cumulant generating function by theorem 4. In fact, by equation 2.6.13, it is the cumulant generating function of a standardized normal distribution, $N(0, 1)$. This, and the uniqueness of the cumulant generating function, proves the result.

The result is truly remarkable because it says that whatever distribution the x_n have in common their sum, suitably standardized, tends to have a normal distribution, only provided the variance exists. We have already met an example: in the simple random walk (§2.5) $x_n = +1$ with probability p and -1 with probability $q = 1-p$; $\sum_{r=1}^{n} x_r$ is the position after n steps of the walk, and was shown, corollary to theorem 2.5.2, to have a limiting normal distribution. The Central Limit Theorem shows that the same limiting normal distribution will hold for any random walk provided only that the steps have a distribution, discrete or continuous, with finite variance. Consequently in many additive processes without barriers it is reasonable to take the distribution of position after a fair number of steps as normal. The proof given in §2.5 of the limiting behaviour of the simple random walk extends without serious difficulty to the general situation, though the proof just given is easier. The fundamental difference equation 2.5.3 becomes in the general discrete case

$$p(s \mid n) = \sum_{r=-\infty}^{\infty} p(s-r \mid n-1)p_r,$$

where $\{p_r\}$ is the common distribution of the steps u_n in equation 2.5.1. In the continuous case

$$p(s \mid n) = \int_{-\infty}^{\infty} p(s-x \mid n-1) f(x) dx,$$

where $f(x)$ is the density of u_n. By arguments similar to those used in §2.5 it is possible to pass in the limit from these more complicated equations to the diffusion equation 2.5.7. We omit the details.

Theorem 1 proves that a Poisson variable, $P(\mu)$, tends to normality as $\mu \to \infty$: precisely, that the distribution function of $(x-\mu)/\sqrt{\mu}$ tends to $\Phi(x)$ as $\mu \to \infty$ when x is $P(\mu)$. This is because if x_i is $P(1)$, $\sum_{i=1}^{n} x_i$ is $P(n)$ (§3.5). Similarly, a Γ-variable, $\Gamma(n, \lambda)$, tends to normality as $n \to \infty$ for fixed λ, because the sum of independent Γ-variables with a common λ is a Γ-variable (§3.5).

The Central Limit Theorem can be generalized beyond the form given here. It is not necessary that the x_n have a common distribution, the distributions can be different provided no one

x_n dominates the rest. For example, if the variance were undefined, because the defining integral or sum diverged, then the variable would be rather dominant. It is not even essential that the x_n be independent, although any dependence must not be so strong that one or a few of the x_n dominate. Essentially the Central Limit Theorem holds provided the summation is a true mixture of a lot of distributions of finite variance in which no component of the mixture is much more important than any other. Thus it is often used to justify saying that errors of measurement (§3.3) are normally distributed. If this were so then the results of §3.3 on means and variances would be of even more value than they are without normality because a normal distribution is completely described by its mean and variance. The argument is that the final error of measurement is the result of many minor errors in the experiment, none of which is itself of importance (if it were the experiment would be re-designed) and therefore the final measurement should be normal. But it is difficult to see why the minor errors are additive. Such empirical evidence as one has suggests some departure from normality, but there are many practical situations in which normal distributions appear to occur; thus in anthropometry men's heights have a normal distribution, though whether this is because a man's height is the sum of a number of contributions from his ancestors is not clear.

One particular situation in which errors of measurement are certainly not normally distributed, is where they are due to rounding-off errors. Suppose a true value μ is measured to the nearest unit giving a value x. Then $x-\mu$ has a distribution confined to the interval $(-\frac{1}{2}, +\frac{1}{2})$ and there is usually no obvious reason for thinking that any one value in this interval is more likely than any other. Consequently it is usually assumed that the density is constant in the interval. That is, rounding-off errors have a uniform distribution (§3.5) in $(-\frac{1}{2}, \frac{1}{2})$ in the units of measurement. Obviously the distribution is far from normal.

The Central Limit Theorem may be expressed in the form of a result about the mean $\bar{x} = n^{-1} \sum_{i=1}^{n} x_i$ of a number of random variables by saying that $(\bar{x}-\mu)/(\sigma/\sqrt{n})$ is approximately normally distributed. The standardizing factors, μ and $\sigma/\sqrt{n}$, are of course the mean and standard deviation of $\bar{x}$. The reduction of random errors by repeated measurement was discussed in §3.3: the Central Limit Theorem shows that the mean of the measurements will not only have a smaller random error than any single measurement but will also have an approximately normal distribution. For the individual measurements will usually be independent and have a common distribution—the assumptions of theorem 1—and only the existence of a variance is needed for the normal limit. Rounding-off errors provide an example. In statistics (that is in inference problems) normality is a most convenient assumption and a statistician often takes advantage of this property of the mean of repeated measurements.

Weak convergence

Theorem 2 is of more theoretical interest. The definition of weak convergence is an attempt to put into precise form the limit notation, 'lim', that was used in chapter 1 as a basis for the probability axioms. In words (1) says that however small a departure from c is considered, the random variable y_n will, for large n, almost always be contained within it. It does not say that y_n tends to c; all it says is that y_n is usually rather near c. We often write $\operatorname{plim}_{n\to\infty} y_n = c$ (probability-limit) to distinguish it from the mathematical limit. The sequence of trials considered in §1.1 was interpreted in §1.3 as a random sequence of trials with constant probability of success. If $x_n = 1$ or 0 according as the nth trial results in a success or a failure, the x_n satisfy the conditions of the law, $\bar{x}$ is the success ratio, m/n, and $\mathscr{E}(x_n) = p = p(x_n = 1)$. Hence the law says that $\operatorname{plim}_{n\to\infty} (m/n) = p$, and is an expression proved within our mathematics corresponding to 'lim' $(m/n) = p$ observed in the real world. Thus the weak law is a confirmation that our mathematics is a good description of random phenomena because it contains within it a result in such good agreement with practice. Had $\operatorname{plim}_{n\to\infty} (m/n)$ not been p it would have been necessary to amend the axioms.

Ergodicity

The law does have some practical consequences. In the language of §3.3, it says that if repeated measurements are made with no systematic error of an unknown μ, then their mean will tend, in the sense of weak convergence, to μ. In other words you will get as near as you wish to the right answer if you go on long enough. Another consequence of the law, and generalizations of it, is a principle that we all use, often without recognizing that it is a result of some depth: this is the *ergodic* principle. In its simple form here it is the statement that the two means, μ and $\operatorname{plim}_{n\to\infty} \bar{x}$, are the same. Hence the mean can be found by taking each sample point or elementary event, a, finding the value of x_1 there, and averaging over all a, so obtaining μ: or one can take a single sample point and find $x_1, x_2, x_3, \ldots$, for this a, and the average of these will be plim $\bar{x}$. Provided the x_i's obey certain conditions the two means will be equal. Ergodic theorems are concerned with conditions for this to be so and the law here is a very simple case. The average over the sample space is often called a *spatial* mean: that over $\{x_n\}$ is called a *temporal* mean since n can be thought of as a measure of time. An example of the use of the ergodic principle arises with the Poisson process. Let x_n be the number of incidents in the interval $(n-1, n)$: clearly the conditions of the law obtain and $\mathscr{E}(x_i) = \lambda$. Then λ may be found either by observing many Poisson processes (all with the same value of λ) in the interval (0, 1) so obtaining $\mathscr{E}(x_1)$, or by observing a single process over a long period of time $(0, n)$ when the average number of incidents per unit time is $\bar{x}$ which tends (in the sense of converges weakly) to λ. Stochastic processes for which the two means are equal are called *ergodic* processes. Examples will be studied in §§4.5, 4.6. Notice that the practical justification for expectation used in §2.1 in connexion with games of chance uses the ergodic principle. Another example is given in §4.2.

Strong convergence

A more useful ergodic result is the strong law of large numbers (theorem 3). A sequence of random variables is said to converge strongly to a constant c if 'almost all' realizations of the sequence converge to c in the ordinary mathematical sense of convergence; where by 'almost all' is meant with probability one. It can be proved that strong convergence implies weak convergence but not conversely, so the nomenclature is reasonable. The strong law, when looked at as an ergodic result, says that the two means, μ and $\lim_{n\to\infty} \sum_{i=1}^{n} x_i/n$, with probability one, are the same and, moreover, because the condition stated is necessary and sufficient, the existence of either mean implies the existence of the other. The probability content of this result is only a qualifying one (namely 'with probability one') and permits the identification of μ and $\lim \bar{x}$ in almost all cases. This fact and the use of an ordinary mathematical limit makes the strong law more useful than the weak one. Also the existence of a simple converse, that if $\lim \bar{x}$ exists, almost always, $\mathscr{E}(x)$ exists and is μ, is of practical value. (A corresponding converse for the weak law is false.) Important applications of the strong law are given in §§4.2 and 4.3. Unfortunately the proof of the strong law involves some manipulations which are outside the scope of this book and we have to leave it. A proof is given in Loève (1960). Notice that the result makes no mention of the existence of $\mathscr{D}^2(x)$. The weak law can also be stated without assuming a variance to exist but the proof given here, which is firmly based on variance considerations, is no longer available. Theorem 4 can be used to provide a proof.

Poisson limit of binomial

An alternative proof of the Poisson limit result (§2.3) that a $B(n, p)$ variable tends, as $n \to \infty$, $p \to 0$ with $np = \mu$, to a $P(\mu)$ variable, is provided by theorem 4. The characteristic function of a $B(n, p)$ variable is obviously $(pe^{it}+q)^n$ and considering the limit of this as $n \to \infty$ under the conditions stated we have

$$\lim_{n\to\infty} \left[1 + \frac{\mu}{n}(e^{it}-1)\right]^n = \exp[\mu(e^{it}-1)]$$

by the well-known result that $\lim_{n\to\infty} (1+[x/n])^n = e^x$. The right-hand side is continuous at $t = 0$ and is recognizable as the characteristic function of $P(\mu)$ (equation 2.6.12, with $z = it$).

Cauchy distribution

We end this chapter with a word of warning. Theorems 1 and 2 are not necessarily true without the restrictions on $\{x_n\}$: normality is not always the end-product of a limiting operation in probability theory. To illustrate we produce the standard skeleton in the statistician's cupboard: the *Cauchy* distribution. This is a continuous distribution with density, for all x,

$$f(x) = \{\pi(1+x^2)\}^{-1}. \tag{5}$$

(The substitution $x = \tan\theta$ shows that the integral of (5) is one.) The integral for the mean does not converge absolutely and hence neither does that for the variance. The characteristic function is $e^{-|t|}$, a result which is easily obtained by complex-variable methods. Hence if the $\{x_n\}$ of theorems 1 or 2 have Cauchy distributions (with no mean nor variance) the sum $\sum_{i=1}^{n} x_i$ has characteristic function $e^{-n|t|}$ and, on substituting t/n for t, we see that the mean, $\sum_{i=1}^{n} x_i/n$, has characteristic function $e^{-|t|}$, the original Cauchy distribution. Hence the mean of measurements with a Cauchy distribution has the same distribution as any single measurement. Thus if the experimenter's measurements had a Cauchy distribution and if he took the average of them he would be in the same position as if he had taken just one and thrown the rest away. The resolution of this apparent paradox is obtained by taking some other function of the measurements in place of the average. For example, the median has a distribution which, as $n \to \infty$, tends to normality with a variance which tends to zero.

Suggestions for further reading

The suggestions listed in chapter 2 are also appropriate for this chapter. Tables of random sampling numbers have been provided by Kendall and Babington Smith (1954). They have been converted, as described in §3.5, to provide tables of random samples from a standardized normal distribution, Wold (1954). Similar collections are given by the Rand Corporation

(1955). The special topic of limit theorems is considered in detail by Gnedenko and Kolmogorov (1954). The multivariate normal distribution is the basis of almost all statistical work involving several random variables. An excellent modern book is Anderson (1958), which is primarily concerned with the statistical aspects. A delightful and unusual book is that by Kac (1959).

Exercises

1. Suppose that 'points' occur in a Poisson process of parameter λ, so that the number, X, of points in a unit interval has a Poisson distribution of mean λ. Suppose, further, that at each point, there is one 'individual' with probability θ and two individuals with probability $1-\theta$, the numbers of individuals associated with different points being mutually independent. Let Y be the number of individuals in a unit interval. By considering the properties of Y conditionally on X, or otherwise, obtain the mean, variance and probability generating function of Y. (Lond. M.Sc.)

2. In n mutually independent trials, where n is even, there is a probability of a 'success' of θ_1 in each of the first $\frac{1}{2}n$ trials and a probability of a success of θ_2 in each of the remaining trials.

Prove that the mean and variance of the total number of successes are respectively $\frac{1}{2}n(\theta_1+\theta_2)$ and $\frac{1}{2}n(\theta_1+\theta_2)-\frac{1}{2}n(\theta_1^2+\theta_2^2)$. Hence show that, unless $\theta_1 = \theta_2$, the variance of the number of successes is less than it would be in a binomial distribution with the same number of trials and the same mean number of successes. (Lond. B.Sc.)

3. A game between two opponents A and B is in five steps, the winner of a step scoring one point. Each step is equally likely to be won by either player and the outcomes of different steps are mutually independent. Prove that the probability that one or other player is ahead throughout the game is 3/8.

What are the mean and variance of the difference, at the end of the game, between the numbers of points scored by A and by B? (Lond. B.Sc.)

4. In a series of mutually independent trials, the possible outcomes on any one trial are 'success' and 'failure'. By introducing a random variable Z_i for the ith trial, equal to 1 if the ith trial is a success and 0 if the ith trial is a failure, or otherwise, prove that the mean and variance of X_n, the number of successes in n trials, are

$$\mathscr{E}(X_n) = n\theta., \quad \mathscr{D}^2(X_n) = n\theta.(1-\theta.)-\Sigma(\theta_i-\theta.)^2.$$

Here θ_i is the probability of a success in the ith trial and $\theta. = \Sigma\theta_i/n$.

What are the values of $p(X_n = 0)$ and $p(X_n = 1)$? (Lond. B.Sc.)

5. Incidents are occurring in a Poisson process at the rate of one per unit of time. Each incident is either of type A with probability p or type B with probability $q = 1-p$: these being independent for the different incidents.

Show that the following two events are equivalent:

(i) at least r out of the first n incidents are of type A;

(ii) the rth incident of type A occurs before the $(n-r+1)$st incident of type B.

Hence show that if $\tilde{s}$ is a $B(n, p)$ variable then

$$p(\tilde{s} \geqslant r) = p(\tilde{u} > \tilde{v}),$$

where $\tilde{u}$ and $\tilde{v}$ are independent random variables having distributions which are respectively $\Gamma(n-r+1, q)$ and $\Gamma(r, p)$.

6. n couples procreate independently with no limits on family size. Births are single and independent and, for the ith couple, the probability of a boy is p_i. The sex ratio, S, is (mean number of all the boys)/(mean number of all the children). Show that if all couples:

(i) do not practice birth control, $S = \Sigma p_i/n$;

(ii) stop procreating on the birth of a boy, $S = n/\Sigma(1/p_i)$;

(iii) stop procreating on the birth of a girl, $S = 1-n/\Sigma(1/q_i)$;

(iv) stop procreating when they have children of both sexes, $S = [\Sigma(1/q_i)-\Sigma p_i]/[\Sigma(1/p_i q_i)-n]$.

Comment on possible inequalities among these ratios.

7. A certain kind of nuclear particle splits into 0, 1, or 2 new particles with probabilities 1/4, 1/2, 1/4 respectively, and then dies. The individual particles act independently of each other. Given a particle, let X_1, X_2 and X_3 denote the number of particles in the first, second and third 'generations' respectively.

Find:

(i) $p(X_2 > 0)$;

(ii) $p(X_1 = 1 \mid X_2 = 1)$;

(iii) $p(X_3 = 0)$.

8. A process of incidents starting from $t = 0$ is such that the probability of an incident in $(t, t+\delta t)$ is $\lambda_n \delta t$, where n is the number of incidents in $(0, t)$, but is otherwise independent of the time of occurrence of the incidents in $(0, t)$. Find the expected time from $t = 0$ up to the nth incident.

9. A game consists of a number of independent turns at each of which three, and only three, mutually exclusive events can occur as follows:

(i) the game terminates without addition to the score;

(ii) the game continues without addition to the score;

(iii) the game continues with the addition of one point to the score.

If a player has a constant probability of encountering these events in each turn (but a different probability for each event), show that the probability of his scoring N points in one game is

$$\frac{m^N}{(1+m)^{N+1}},$$

where m is his average score per game.

Deduce that the probability of a player with average m_1 scoring more points in one game than a player with average m_2 is

$$\frac{m_1}{1+m_1+m_2}.$$

10. Each of $N+1$ boxes contains N counters. The boxes are numbered $i = 0, 1, 2, \ldots, N$; box i has i red counters and $(N-i)$ green counters well mixed. Box k is arbitrarily chosen by an investigator, and from it he makes N independent drawings of a counter *with replacement* obtaining a red counter r times. He then makes n drawings *without replacement* from box number r, where n is a fixed number $\leqslant N$, and obtains s red counters. Find the frequency distribution of s under repetitions of the sampling with fixed N, k and n. Discuss the manner in which the mean value of s depends on k. (Aberdeen Dip.)

11. Obtain the regression equations of y on x for the bivariate probability distribution where $p(x=r, y=s)$, r and s integers, is the coefficient of $u^r v^s$ in

$$\exp\,[a(u-1)+b(v-1)+c(u-1)\,(v-1)].$$

(Aberdeen Dip.)

12. The densities $\lambda_1, \lambda_2, \ldots$ of different species of nocturnal insects in the neighbourhood of a light-trap are supposed to be observations of a random variable λ which has a moment generating function of the form

$$\mathscr{E}(e^{\lambda t}) = \{f(t)\}^{\alpha},$$

where α is a constant, $\alpha \geqslant 0$. The number caught of the ith species is supposed to be a Poisson variable with mean λ_i. If r is the number caught of a randomly chosen species find the probability generating function of r. What form does this take, *conditional on* $r \geqslant 1$? Obtain the limiting form of this conditional probability generating function as $\alpha \to 0$. Hence, or otherwise, obtain the distribution of frequency of individuals among species caught in the trap when the probability density function of λ is

$$\frac{\lambda^{\alpha-1}e^{-\lambda}}{\Gamma(\alpha)} \quad (\lambda > 0),$$

and the number of species present in the vicinity of the trap is very large compared with the number of different species represented in the trap.

(Lond. Dip.)

13. Producer gas from a certain plant has calorific value varying erratically in a normal distribution of mean 160 Btu per cu.ft and standard deviation 28 Btu per cu.ft. It is desired that the calorific value shall only rarely fall below 110 Btu per cu.ft. Find the proportion of values actually below this limit.

A new flow of gas is formed by mixing two similar streams, the calorific values of which vary independently in normal distributions with the above mean and standard deviation. The calorific value of the combined stream at any point is the average of the corresponding calorific values in the separate streams. Find the proportion of values in the new flow below 110 Btu per cu.ft. (Camb. N.S.)

14. At one stage in the manufacture of an article a piston of circular cross-section has to fit into a similarly shaped cylinder. The distributions of diameters of pistons and cylinders are known to be normal with parameters:

Piston diameters: mean 10·42 cm, standard deviation 0·03 cm.

Cylinder diameters: mean 10·52 cm, standard deviation 0·04 cm.

If the pistons and cylinders are selected at random for assembly, what proportion of pistons will not fit into the cylinders? What is the chance that in 100 pairs, selected at random, all the pistons will fit? (Wales Math.)

15. The variates y_1, y_2, have a bivariate normal distribution with means zero, variances σ_1^2, σ_2^2 and covariance $\rho\sigma_1\sigma_2$. Individuals are selected at random subject to the single constraint

$$y_1 \geqslant x\sigma_1.$$

For this selected portion of the population find: (i) $\mathscr{E}(y_1)$, (ii) $\mathscr{E}(y_2)$, (iii) $\mathscr{D}^2(y_1)$, (iv) $\mathscr{D}^2(y_2)$, (v) $\mathscr{C}(y_1, y_2)$, in terms of x, z, p, where

$$z = \frac{1}{\sqrt{(2\pi)}}\, e^{-\frac{1}{2}x^2} \quad \text{and} \quad p = \int_{-\infty}^{x} \frac{1}{\sqrt{(2\pi)}}\, e^{-\frac{1}{2}t^2}\, dt.$$

(Aberdeen Dip.)

16. Obtain the regression equations of y on x for the bivariate probability distribution

$$f(x, y)\, dx\, dy = \frac{y \exp\left[-y/(1+x)\right] dx\, dy}{(1+x)^4}$$

for $x, y \geqslant 0$. (Aberdeen Dip.)

17. In a certain investigation the frequency of accidents per unit time for a given individual follows a Poisson distribution with expectation m, but the value of m varies from one worker to another in a distribution with frequency function

$$\frac{\alpha^\lambda}{\Gamma(\lambda)}\, m^{\lambda-1} e^{-\alpha m},$$

where α and λ are constants. Show that in a random sample of workers from this population the frequency of 0, 1, 2, ... accidents per unit time would be represented by the terms in an expansion of the negative binomial $(q-p)^{-\lambda}$, where $q-p = 1$. Express p in terms of α and λ. (Camb. Trip.)

18. In a sequence of independent trials, a certain event E has always the same chance p of occurring $(0 < p < 1)$. Let r be the number of the trial at which E first occurs. Find the probability distribution of r, and show that the mean and variance are respectively

$$\frac{1}{p} \quad \text{and} \quad \frac{1-p}{p^2}.$$

A large population consists of equal numbers of individuals of c different types. Individuals are drawn at random one by one until at least one individual of each type has been found, whereupon sampling ceases. Show that the mean number of individuals in the sample is

$$c\left\{1+\frac{1}{2}+\frac{1}{3}+\ldots+\frac{1}{c}\right\}$$

and the variance of the number is

$$c^2\left\{1+\frac{1}{2^2}+\frac{1}{3^2}+\ldots+\frac{1}{c^2}\right\}-c\left\{1+\frac{1}{2}+\frac{1}{3}+\ldots+\frac{1}{c}\right\}.$$

(Camb. N.S.)

19. The random variables $X_1, X_2, \ldots, X_n$ (which are not necessarily independent) are such that each takes only the values 0 or 1. The probability that $X_i = 1$ is p_i and the probability that $X_i = 1$, $X_j = 1$ simultaneously is p_{ij}. Calculate the expectation and variance of X_i and the covariance of X_i and X_j. If $nY = \sum_{i=1}^{n} X_i$, show that the variance of Y is

$$(S_2 - S_1^2) + (S_1 - S_2)/n,$$

where $$nS_1 = \sum_{i=1}^{n} p_i \quad \text{and} \quad \tfrac{1}{2}n(n-1)S_2 = \sum_{j=2}^{n}\sum_{i=1}^{j-1} p_{ij}.$$

When the variables are independent show that the variance of Y can be expressed entirely in terms of n, S_1 and V, where $nV = \sum_{i=1}^{n}(p_i - S_1)^2$. Deduce the variance of the number of successes in n independent trials when the probability of success per trial is p. (Camb. N.S.)

20. If x and y are independent, find the mean and variance of

$$axy+bx+cy+d$$

in terms of the means and variances of x and y, where a, b, c and d are constants.

The length x and breadth y of manufactured rectangular plates are found to form two independent series of measurements with means 16 and 25 cm and standard deviations 1 and 2 cm respectively. Assuming that the area of the plates has a normal distribution, find what proportion of the plates will have an area greater than 480 cm². (Camb. N.S.)

21. If x and y are independent show that

$$\mathscr{D}^2(xy) = \mathscr{D}^2(x)\,\mathscr{D}^2(y) + [\mathscr{E}(x)]^2\,\mathscr{D}^2(y) + [\mathscr{E}(y)]^2\,\mathscr{D}^2(x).$$

22. Solid right circular metal cylinders of height h and radius r are accepted on testing after manufacture provided that their volumes do not exceed 57,500 c.c. The dimensions h and r are distributed independently with the same standard deviation ρ about the respective mean values 20 and 30 cm. Find the greatest tolerable value for ρ if the frequency of rejection is not to exceed $2\frac{1}{2}$%, it being assumed that the logarithm of the volume is distributed approximately normally. (Camb. N.S.)

23. It is required to estimate the mean percentage yield of a certain ore. To do this, r separate specimens of ore are taken at random, and s independent laboratory determinations are made on each specimen. It is known that the variance of the true percentage yield of different specimens is σ_1^2, whilst each laboratory determination is, in addition, subject to an independent experimental error of variance σ_0^2.

Find the variance σ^2 of the estimated mean percentage yield obtained by averaging the rs determinations.

If it takes time T_1 to collect a specimen and T_0 to make a determination, find the values of r and s which minimize the time necessary to attain a given value of σ^2. (Camb. N.S.)

24. The number, n, of particles recorded by a Geiger counter in an interval of time T when placed near a weak source of radiation has a Poisson distribution with mean λT, where λ measures the strength of the source. Show that $\mathscr{E}(n/T) = \lambda$ and find $\mathscr{D}^2(n/T)$.

The source of radiation is only available in the presence of an independent noise radiation of constant strength μ, unknown, and in order to measure λ a second method is used where the number, n_1, of particles is recorded for an interval T_1 with only the noise present, and then the number n_2 for an interval T, with both noise and source present. Show that

$$\mathscr{E}(n_2/T_2 - n_1/T_1) = \lambda$$

and find $\mathscr{D}^2(n_2/T_2 - n_1/T_1)$.

Show that, in order to make the two variances equal, if the noise and source are found to be about the same strength and $T_1 = T_2$, then it requires the counter to be used for about six times as long when the noise is present, as would be needed if the noise were absent and the first method could be used. Can the factor six be reduced in these circumstances by allowing T_1 and T_2 to be unequal? (Camb. N.S.)

25. The distance r from the centre of a plane target to the point where a shot falls has a probability distribution of the form

$$\frac{1}{\sigma^2} e^{-r^2/2\sigma^2} r\,dr \quad (0 < r < \infty),$$

where σ is initially unknown. In order to estimate the probability P that a shot will fall within a fixed distance a of the centre of the target, n shots are fired and the distances $r_1, r_2, \ldots, r_n$ of their points of fall from the centre are recorded. P is then estimated as

$$\hat{P} = 1 - e^{-a^2/2s^2},$$

where

$$s^2 = \frac{1}{2n} \sum_{i=1}^{n} r_i^2.$$

Show that, if n is large, $\hat{P}$ has mean approximately equal to P and variance approximately equal to

$$\frac{a^4}{4n\sigma^4} e^{-a^2/\sigma^2}.$$

Can you suggest any alternative method by which P might be estimated after the n shots have been fired? (Camb. N.S.)

26. The number of insects killed, r, when n insects are exposed to a dose of an insecticide, is distributed in a binomial distribution with parameters n and p. The following quantity y is defined in terms of r and n:

$$y = \ln \frac{2r+1}{2n-2r+1}.$$

By writing $r = np+\delta r$ and expanding y as a Taylor series in ascending powers of δr, or otherwise, show that if n is large the distribution of y has approximately mean $\ln \{p/(1-p)\}$ and variance $1/\{np(1-p)\}$, where terms of order n^{-2} have been neglected.

The chance p that any insect is killed depends on the dose, according to a relation of the form

$$\ln \frac{p}{1-p} = \alpha+\beta x,$$

where x is the logarithm of the dose, and α and β are unknown constants. For the purpose of estimating α and β, n insects are tested independently at each of the two doses for which $x = 0$ and 1 respectively. The numbers of insects killed are found to be respectively r_0 and r_1. If the corresponding values of y are denoted by y_0 and y_1, then y_0 and y_1-y_0 are taken as estimates of α and β respectively. Find expressions for the variances and the covariance of these two estimates, in terms of α and β, using the above approximate expression for the variance of y. (Camb. N.S.)

27. In attempting to construct an equilateral triangle of unit side correlated errors each having zero mean are made in each of the three sides x_1, x_2, x_3. The error in each side has variance σ^2 and there is a covariance $(-\rho\sigma^2)$ between each pair of errors, $\rho > -\frac{1}{2}$. Show that the area of the triangle is distributed with variance given approximately by $\sigma^2(1-2\rho)/4$. (Camb. N.S.)

28. The random variable r has a *negative binomial distribution*, that is,

$$p_r = \binom{r-1}{n-1} p^n(1-p)^{r-n} \quad (r = n, n+1, \ldots),$$

where n and p are parameters of the distribution. Obtain a transformation $f(r/n)$ of r/n such that the variance of the transformed variable is approximately independent of p. (Camb. N.S.)

29. In a certain family of distributions encountered in field entomology a variate x is distributed with mean m and variance $\mu(m+\lambda m^2)$ where λ and μ are the same for all distributions of the family and μ is small. It is desired to apply a functional transformation $f(x)$ which will yield a variate y whose variance V is approximately independent of m. Determine appropriate forms for $f(x)$ in the cases (i) $\lambda > 0$, (ii) $\lambda = 0$. (Camb. N.S.)

30. x is a Poisson variable with expected value ξ. y is equal to $(x+a)^b$ where a is a non-negative quantity. Show that:

(i) $\mathscr{D}^2(y) = b^2(\xi+a)^{2b}\,[\xi^{-1}+\{\frac{3}{2}(b-1)^2-2a\}\,\xi^{-2}+O(\xi^{-3})]$,

(ii) $\sqrt{[\beta_1(y)]} = (3b-2)\xi^{-\frac{1}{2}}+O(\xi^{-\frac{3}{2}})$.

Using these formulae, suggest values of b and a to be used to produce a variable y with variance which changes with ξ as little as possible (relative to its magnitude). What further information would be needed if the aim were to make $\sqrt{[\beta_1(y)]}$ as near as possible to zero? $[\beta_1 = \mu_3^2/\mu_2^3.]$ (Lond. Dip.)

31. The height h of a vertical tower whose base is at a horizontal distance 500 m from the point of observation is determined by measuring E the angle of elevation of the top of the tower. If the measurement of E is distributed with standard deviation 0·1° about the mean value 3·6°, find approximations to the mean and standard deviation of the value obtained for h. (Camb. N.S.)

32. x has probability density

$$\frac{1}{n!}x^n e^{-x}, \quad \text{for} \quad x \geqslant 0,$$

and zero otherwise. Show that $\sqrt{x}$ has a standard deviation which, to a first approximation, does not depend on n, and determine its value. (Camb. N.S.)

33. The co-ordinates x, y of a point whose true position is P are measured with variances σ_x^2, σ_y^2 respectively, the estimated position being P'. Show that the mean of the square of the radial error PP' is $\sigma_x^2+\sigma_y^2$, whether or not the errors in x and y are independent.

A landmark at P, whose co-ordinates x, y are unknown, is observed by radar from two known points, P_1, P_2; the distances PP_1 and PP_2 being measured independently with standard deviation σ. If σ is small compared with PP_1 and PP_2, show that the root mean square deviation of the radial error in the estimated position of P is approximately $\sqrt{2}\sigma\operatorname{cosec}\theta$, where θ is the angle $P_1\hat{P}P_2$. (It may be assumed that P is known to be on one side of P_1P_2, and that θ is not nearly 180°.) (Camb. N.S.)

34. The height, h, of a hill above sea-level is estimated by two observers at sea-level independently measuring its angular elevation from two points at a known distance, s, apart on the same bearing from the summit. Find the approximate variance in h corresponding to small errors in the measurements of the angles. If one observer is known to be more accurate than the other, which should be placed nearer the hill in order to minimize the variance of h? (Wales Math.)

35. A sample $x_1, x_2, \ldots, x_n$ of independent observations is drawn from a population in which x is distributed normally with zero mean value and standard deviation σ. Find the mean and variance of $|x|$ and x^2.

Deduce that

$$s_1 = \frac{1}{n}\sqrt{\frac{\pi}{2}}\sum_{k=1}^{n}|x_k|$$

has mean value σ and variance $V = (\frac{1}{2}\pi-1)\sigma^2/n$, and that $s_2^2 = \sum_{k=1}^{n}(x_k^2/n)$ has mean value σ^2 and variance $2\sigma^4/n$. (Camb. N.S.)

36. Neutrinos are emitted from an atomic pile independently and at random at an average rate of one per unit time. If x is the interval between the first and the second and y is the interval between the second and the third, find the joint probability density of u and v, where $u = \sqrt{(xy)}$ and $v = \sqrt{(y/x)}$. Find the probability density of v and express that of u in terms of $K_0(2u)$, where

$$K_0(z) \quad \text{is} \quad \int_0^\infty e^{-z\cosh\theta}\,d\theta. \qquad \text{(Camb. N.S.)}$$

37. Two resistors are each to be constructed by connecting in series ten coils of wire chosen at random from a large collection of coils in which the electrical resistance is a normal variate with a standard deviation of 0·1 ohm. Find the probability that the completed resistors will differ in resistance by more than 1 ohm. (Camb. N.S.)

38. If x and y are two correlated normal variables whose joint probability distribution is given by

$$dF = \frac{1}{2\pi\sigma^2\sqrt{(1-\rho^2)}}\exp\left[-\frac{1}{2\sigma^2(1-\rho^2)}(x^2-2\rho xy+y^2)\right]dx\,dy,$$

show that the transformed variates

$$u = (x+y)/\sqrt{2}, \quad v = (-x+y)/\sqrt{2},$$

are uncorrelated. Hence, or otherwise, show that any curve in the x, y plane on which the probability density is constant is an ellipse with eccentricity $\sqrt{\{2\rho/(1+\rho)\}}$.

What is the probability that a pair of values (x, y) chosen at random will be represented by a point inside the ellipse of constant probability density with area A? (Camb. N.S.)

39. The variables x and y are distributed independently each according to $N(0, 1)$. Show that the joint distribution of $X = x^2$ and $Y = y^2$ is

$$\frac{1}{2\pi}e^{-\frac{1}{2}(X+Y)}X^{-\frac{1}{2}}Y^{-\frac{1}{2}}dX\,dY \quad (0 \leqslant X < \infty, 0 \leqslant Y < \infty).$$

Making the transformation $X = U$, $Y = UV$, show that the joint distribution of U and V is

$$\frac{1}{2\pi}e^{-\frac{1}{2}U(1+V)}V^{-\frac{1}{2}}dU\,dV \quad (0 \leqslant U < \infty, 0 \leqslant V < \infty).$$

Deduce that the ratio $V = Y/X$ is distributed as

$$\frac{1}{\pi}\frac{dV}{(1+V)\sqrt{V}} \quad (0 \leqslant V < \infty). \qquad \text{(Camb. N.S.)}$$

40. If $x_1, x_2, \ldots, x_n$ are independent $N(\mu, \sigma^2)$ variables show that the variables

$$\sqrt{n}\bar{x} = (x_1+x_2+\ldots+x_n)/\sqrt{n},$$
$$y_1 = (x_1-x_2)/\sqrt{2},$$
$$y_2 = (x_1+x_2-2x_3)/\sqrt{6},$$
$$\ldots$$
$$y_{n-1} = (x_1+x_2+\ldots-(n-1)x_n)/\sqrt{[n(n-1)]},$$

are also independently and normally distributed with variance σ^2.

Prove that

$$\Sigma(x_i-\mu)^2 = n(\bar{x}-\mu)^2+\Sigma y_i^2,$$

and hence show that $\bar{x}$ and $\Sigma(x_i-\bar{x})^2$ are independently distributed according to $N(\mu, \sigma^2/n)$ and $\Gamma(\frac{1}{2}(n-1), 1/2\sigma^2)$ respectively.

41. Show that if $x_1, x_2, \ldots, x_n$ have a multivariate normal distribution with dispersion matrix **C** then the random variables

$$y_i = \sum_{j=1}^{n} t_{ij}x_j$$

also have a multivariate normal distribution with dispersion matrix **T′CT**, where **T** is the matrix of elements t_{ij}.

In particular show that if the x's are independent and **T** is orthogonal then the y's are independent.

42. If $x_1, x_2, \ldots, x_n$ are independently each distributed as $N(0, \sigma^2)$ show that $\sum_{i=1}^{n} x_i^2$ has a $\Gamma(\frac{1}{2}n, 1/2\sigma^2)$ distribution.

43. The random variables $X_1, X_2, \ldots, X_n$ are independently normally distributed with zero mean and unit variance.

Prove that the density of $Z = X_1^2+X_2^2-X_3^2-X_4^2$ is $\frac{1}{4}e^{-\frac{1}{2}|z|}$.

Obtain the cumulants of

$$U = \sum_{j=1}^{n} jX_j^2. \qquad \text{(Lond. M.Sc.)}$$

44. Let $X_1, X_2, \ldots, X_n$ be independently normally distributed with zero mean and unit variance and let $Y = X_1^2+X_2^2+\ldots+X_n^2$. Obtain the probability density function, mean and variance of $\log Y$. (Lond. M.Sc.)

45. The random variables $X_1, X_2, \ldots, X_n$ are independently distributed with probability density function

$$\frac{e^x}{(1+e^x)^2} \quad (-\infty < x < \infty).$$

The random variable $X_{(i)}$ is the ith largest of $X_1, \ldots, X_n$ when arranged in ascending order of magnitude. Prove that the moment generating function of $X_{(i)}$, $\mathscr{E}(e^{sX_{(i)}})$, is equal to

$$\frac{\Gamma(i+s)\,\Gamma(n-i-s+1)}{\Gamma(i)\,\Gamma(n-i+1)}.$$

Hence, or otherwise, show that for $i-1 > n-i$

$$\mathscr{E}(X_{(i)}) = \sum_{r=n-i+1}^{i-1} \frac{1}{r}. \qquad \text{(Lond. M.Sc.)}$$

46. The bivariate random variable $\mathbf{x} \equiv (x_1, x_2)$ has probability density function

$$p(\mathbf{x}) = Ke^{-\frac{1}{2}Q} \quad (-\infty < x_1 < \infty, -\infty < x_2 < \infty),$$

with

$$Q \equiv (\mathbf{x}-\boldsymbol{\mu})\,\mathbf{A}(\mathbf{x}'-\boldsymbol{\mu}'),$$

where $\boldsymbol{\mu} \equiv (\mu_1, \mu_2)$ is a vector of constants, $\mathbf{A}$ is a symmetric positive definite 2×2 matrix and K is a constant. Find the characteristic function of $\mathbf{x}$ and hence or otherwise evaluate K and $\mathbf{A}$ in terms of the second-order moments of $\mathbf{x}$. Prove that $\frac{1}{2}Q$ is exponentially distributed. If $\mathbf{x}^{(1)}, \mathbf{x}^{(2)}, \ldots, \mathbf{x}^{(n)}$ is a sample of n random and independent observations of $\mathbf{x}$ and if R is the largest of the n corresponding values of $\frac{1}{2}Q$ find the characteristic function of R and hence or otherwise show

$$\mathscr{E}\{R\} = 1+\frac{1}{2}+ \ldots +\frac{1}{n}, \quad \mathscr{D}^2\{R\} = \frac{\pi^2}{6}.$$

(Lond. Dip.)

47. The random variables X_1, X_2, X_3 are independently distributed with probability density function

$$\lambda e^{-\lambda x} \quad (x \geqslant 0),$$
$$0 \quad (x < 0).$$

Show that their range w has density

$$2\lambda e^{-\lambda w}(1-e^{-\lambda w}) \quad (w \geqslant 0),$$
$$0 \quad (w < 0).$$

Hence show that the ratio of the mean range to the standard deviation of the X's, which may be taken without proof to be $1/\lambda$, is $3/2$.

(Lond. B.Sc.)

48. X and Y are normally and independently distributed with zero mean and unit variance, and

$$Z = \rho X + \sqrt{(1-\rho^2)}\, Y.$$

Show that the joint density of X and Z is a bivariate normal distribution with correlation coefficient ρ. Hence explain how, given an ordinary table of random digits, you would construct a random sample of observations from the bivariate normal distribution with zero means, unit standard deviations and correlation coefficient 0·6. (Lond. B.Sc.)

49. A random variable X is equally likely to assume any value in the interval 0 to θ and cannot be outside this interval. A sample of n independent values of X are available, of which $X_{(1)}$ is the lowest and $X_{(n)}$ the highest. Show that the joint probability density function of $X_{(1)}$ and $X_{(n)}$ is

$$f(x_1, x_n) = n(n-1)\,(x_n - x_1)^{n-2}/\theta^n \quad (0 \leqslant x_1 \leqslant x_n \leqslant \theta).$$

Hence, or otherwise, show that if $W = X_{(n)} - X_{(1)}$ is the sample range, its density is

$$\frac{n(n-1)\,w^{n-2}}{\theta^{n-1}}\left(1-\frac{w}{\theta}\right) \quad (0 < w \leqslant \theta).$$

(Lond. B.Sc.)

50. Metal bars have flaws distributed randomly along their length at rate λ, i.e. the probability that there is a flaw between $(x, x+\delta x)$, where δx is very small, is $\lambda\delta x$ independently of the positions of other flaws. Prove that the probability that a rod of length l contains no flaws is $e^{-\lambda l}$.

Conditionally on there being no flaws, the strength of a rod is distributed with mean μ_0 and variance σ_0^2. Conditionally on there being one or more flaws, the strength is distributed with mean μ_1 and variance σ_1^2. Prove that the mean and variance of strength are respectively

$$\mu_0 e^{-\lambda l} + \mu_1(1-e^{-\lambda l})$$

and

$$\sigma_0^2 e^{-\lambda l} + \sigma_1^2(1-e^{-\lambda l}) + (\mu_1-\mu_0)^2 e^{-\lambda l}(1-e^{-\lambda l}).$$

(Lond. B.Sc.)

51. Show that if x_1 and x_2 are independently and uniformly distributed over the interval a to b, then the probability density function of $u = x_1+x_2$ is given by

$$\begin{aligned} f_u(u) &= 0 \quad \text{for} \quad u < 2a \text{ or } u > 2b \\ &= \frac{u-2a}{(b-a)^2} \quad \text{for} \quad 2a < u < a+b \\ &= \frac{2b-u}{(b-a)^2} \quad \text{for} \quad a+b < u < 2b. \end{aligned}$$

52. The length AB has an exponential distribution with parameter $\frac{1}{2}$. Let O be the mid-point of AB and let C be the intersection of a randomly drawn line from O with the semicircle on AB with centre O. Show that if D is the projection of C on AB, then AD and DB are independently distributed with Γ-distributions, and that if we put

$$\begin{aligned} z &= +(AD)^{\frac{1}{2}} \text{ with probability } \tfrac{1}{2}, \\ &= -(AD)^{\frac{1}{2}} \text{ with probability } \tfrac{1}{2}, \end{aligned}$$

then z has a standardized normal distribution.

53. Show that the minimum of k independent exponential random variables with parameter λ is itself exponential with parameter $k\lambda$.

54. Let $T_1, T_2, \ldots$ be the winning times recorded in successive years for a certain athletic event. T_k is the 'first record' if T_k is the smallest of $T_1, \ldots, T_k$ and T_1 is the next-smallest. Assuming that the r.v.'s T_i are independently and identically distributed and $p(T_i = T_j) = 0$ for $i \neq j$, show that

(i) $p(k \geqslant k_0) = 1/(k_0-1)$,
(ii) $\mathscr{E}(k)$ is infinite.

55. A random variable has distribution function $F(x)$. Show that the largest of n independent observations has distribution function $F^n(x)$, and hence derive also the density function (if this exists).

Three independent observations are taken when the density is

$$\tfrac{1}{2}\exp(-|x|) \quad \text{for all } x.$$

What is the expected value of the largest observation in this case?

56. A pair of random variables is said to have a circular distribution if their joint probability density $f(x, y)$ depends only on x^2+y^2. Show that the joint characteristic function $\phi(t, u)$ of such a pair depends only on t^2+u^2.

Hence, or otherwise, show that, if a pair of independent random variables has a circular distribution, then they must be normally distributed with zero means and equal variances. (Camb. Trip.)

57. The random variables $x_1, x_2, \ldots, x_n$ have a multivariate normal distribution with zero means and dispersion matrix **V**. If **x** is the row-vector $(x_1, x_2, \ldots, x_n)$, prove that two linear forms $\mathbf{a}'\mathbf{x}$ and $\mathbf{b}'\mathbf{x}$ are independent iff $\mathbf{a}'\mathbf{V}\mathbf{b} = 0$. (Camb. Trip.)

58. The variables x_1, x_2, x_3 are each $N(0, 1)$ and independent. Show that $r = (x_1^2+x_2^2+x_3^2)^{\frac{1}{2}}$ and $u = x_1/r$ are independently distributed, and that u is equally likely to lie anywhere in the interval $-1 \leqslant u \leqslant 1$. (Camb. N.S.)

59. The random variables $X_1, X_2, \ldots$ have mean zero, equal finite variance σ^2 and the correlation coefficient between X_i and X_{i+1} is ρ (for all i) and zero for all other pairs of variables. State and prove the weak law of large numbers for $(X_1 + \ldots + X_n)/n$. (Lond. Dip.)

60. Let $x_1, x_2, \ldots, x_n$ be independent random variables each assuming the values $0, 1, 2, \ldots, a-1$ with probabilities $1/a$. Let $s_n = x_1 + \ldots + x_n$. Show that the probability generating function of s_n is

$$\left[\frac{1-z^a}{a(1-z)}\right]^n$$

and hence that

$$p(s_n = j) = \frac{1}{a^n} \sum_{r\geqslant 0} (-1)^r \binom{n}{r} \binom{n+j-ar-1}{j-ar},$$

where only finitely many terms in the sum are different from zero.

61. Let Z_j $(j = 0, 1, 2, \ldots)$ be a sequence of independent random variables which take only the values 0 and 1. Let the probability that Z_j takes the value 1 be p_j. If $\sum_{j\geqslant 0} p_j$ diverges, show that there is probability 1 that $\sum_{j\geqslant 0} Z_j$ diverges. What happens when $\sum_{j\geqslant 0} p_j$ converges?

(Camb. Trip.)

62. A certain type of plant lives for precisely one year. Each autumn it produces seeds, the plants from which produce seed in the following autumn.

If $F_r(x)$ is the probability generating function for the number of rth generation descendants of a given plant, show that

$$F_{r+s}(X) = F_r\{F_s(x)\}$$

for any integers $r > 0$, $s > 0$.

Let $G_r(x)$ be the probability generating function for the total number of descendants of a given plant of at most the rth generation, excluding the original plant. Show that

$$G_1(x) = F_1(x),$$

and

$$G_{s+1}(x) = F_1\{xG_s(x)\}$$

for any integer $s \geqslant 1$.

Let the expectation and variance of the number of first-generation descendants be k and v, respectively. Show that the expectation of the total number of descendants of a given plant is finite if and only if $k < 1$, and that then the variance is $v/(1-k)^3$. (Camb. Trip.)

63. If an animal has a chance p of being female and q of being male (where $p+q = 1$), independent of other animals in the litter, find the probability of being able to make at least one mating of a male and a female from a litter of s animals. If, for $s \geqslant 1$, the probability of getting a litter of size s is proportional to the sth term of the Poisson series, prove that the proportion of litters from which matings can be made is

$$(1-e^{-pm})\,(1-e^{-qm})/(1-e^{-m}),$$

where m is the mean litter size.

4

STOCHASTIC PROCESSES

The definitions and theorems of the first three chapters are here applied to the study of processes developing in time according to probabilistic laws. Two such stochastic processes have already been introduced: the Poisson process (§2.3) and the Random Walk (§2.5); and the results for these are basic in the study of the more elaborate processes of this chapter. In the first three sections particular processes are discussed and serve to introduce several important ideas. In the fourth section a branch of theory is developed and applied to a special type of process in the remaining two sections. The material of this chapter is not used in the remainder of the book. Readers interested in statistics may proceed directly to chapter 5.

4.1. Immigration–emigration process

It is convenient to prove here a result, the sufficiency part of which we shall need in this section. The necessity part will be used in §4.3.

Theorem 1. *If x is a positive random variable with $p(x > t) > 0$ for all t, then the necessary and sufficient condition that*

$$p(x \leqslant t+u \mid x > t)$$

does not depend on t, for any positive u, is that x has an exponential distribution.

The conditional probability may be written in terms of $G(t) = p(x > t)$,

$$p(x \leqslant t+u \mid x > t) = \frac{p(t < x \leqslant t+u)}{p(x > t)} = \frac{p(x > t) - p(x > t+u)}{p(x > t)}$$

$$= \{G(t) - G(t+u)\}/G(t).$$

This is to depend only on u, so we may write it $H(u)$, say. Then

$$G(t) - G(t+u) = G(t)\, H(u).$$

If $t = 0$, $G(0) = 1$, since the random variable is positive. Hence $1-G(u) = H(u)$ and, eliminating $H(u)$,

$$G(t)-G(t+u) = G(t)\,[1-G(u)],$$

or

$$G(t+u) = G(t)\,G(u),$$

whence $G(t) = e^{-\kappa t}$ for some $\kappa > 0$ (cf. equation 2.3.1). Since $G(t)$ is one minus the distribution function of x, the result is proved.

Consider a Poisson process of rate λ in which the incidents are arrivals of particles in a region. A particle stays in the region for a time t, which is a random variable with an exponential distribution of parameter κ, $E(\kappa)$. The times of stay for different particles are independent and independent of the incoming Poisson process. We wish to study the distribution of the number of particles in the region and how it changes with time.

If x is the time of stay in the region of a particle, the probability studied in theorem 1 is the probability that the particle will leave in $(t, t+u)$ given that it was there at time t. Since x is $E(\kappa)$, the sufficiency part of the theorem shows that this probability does not depend on t. In other words, the chance that it will leave in an interval of length u does not depend on how long it has been there. The theorem allows us to complete the specification of the *immigration–emigration process* by supposing that at the start, $t = 0$, there were N particles in the region—it is not necessary to say how long these N particles had been there.

The argument used to derive the required distribution of the number in the region is based on relating the number at time t (thought of as 'now') with the number at $t-\delta t$ (thought of as the immediate 'past') where δt is small and ultimately tends to zero (cf. example 2.6.3). We saw in §2.3 that the Poisson process had the property that the probability of an incident in any interval of length δt was $\lambda\,\delta t$, and of more than one incident was small, technically $o(\delta t)$, irrespective of the behaviour prior to the interval. Theorem 1 with $u = \delta t$ shows that the probability of a departure of a given particle in any interval of length δt is $\kappa\,\delta t$, and therefore of more than one departure is small, irrespective of the behaviour prior to the interval. These two facts enable the situation at t to be simply related to that at $t-\delta t$.

If there are n particles in the region the probability that an unspecified one of them will leave in δt is, by the binomial distribution and the independence of the particles,

$$\binom{n}{1}\kappa\,\delta t(1-\kappa\,\delta t)^{n-1} = n\kappa\,\delta t,$$

ignoring terms that are $o(\delta t)$, as we will throughout the rest of the argument.

If there are n particles in the region at time t, then the position at time $t-\delta t$ could have been:

(i) $(n-1)$ in the region and one arrived in $(t-\delta t, t)$: the arrival is an event of probability $\lambda\,\delta t$ (if $n = 0$ this possibility can be ignored).

(ii) $(n+1)$ in the region and one left in $(t-\delta t, t)$: the departure is an event of probability $(n+1)\,\kappa\,\delta t$ by the argument at the top of the page.

(iii) n in the region and none left nor arrived in $(t-\delta t, t)$: the latter event has probability $1-\lambda\,\delta t - n\kappa\,\delta t$, since the probabilities must add to one. All other possibilities have probabilities $o(\delta t)$ which can be ignored. Hence if

$$p_n(t) = p(n \text{ in region at } t \,|\, N \text{ in region at } t = 0)$$

the generalized addition law gives

$$p_n(t) = p_{n-1}(t-\delta t)\,\lambda\,\delta t + p_{n+1}(t-\delta t)\,(n+1)\,\kappa\,\delta t \\ + p_n(t-\delta t)\,(1-\lambda\,\delta t - n\kappa\,\delta t). \quad (1)$$

This holds for $n > 0$: if $n = 0$ it is still true provided $p_{-1}(t-\delta t)$ is treated as zero. Subtract $p_n(t-\delta t)$ from both sides of (1) so that the right-hand side contains only terms of order δt, divide throughout by δt and let $\delta t \to 0$: the result is

$$dp_n(t)/dt = \lambda p_{n-1}(t) + \kappa(n+1)\,p_{n+1}(t) - (\lambda + n\kappa)\,p_n(t). \quad (2)$$

This equation is most easily solved by using the probability generating function (§2.6) of n for fixed t, namely

$$\Pi(x, t) = \sum_{n=0}^{\infty} p_n(t)\,x^n \quad (3)$$

(cf. example 2.6.3). A differentiation of (3) gives

$$\partial\Pi(x, t)/\partial x = \sum_{n=1}^{\infty} np_n(t)\,x^{n-1}, \quad (4)$$

so that if (2) is multiplied by x^n and summed over all n the result is

$$\partial\Pi/\partial t = \lambda(x-1)\,\Pi - \kappa(x-1)\,\partial\Pi/\partial x. \tag{5}$$

This is a partial differential equation of a standard type and may be solved by the usual methods, but a simpler solution is provided by the substitutions of

$$\left.\begin{aligned} s &= (x-1)\,e^{-\kappa t} \\ \text{and}\qquad z &= \lambda x/\kappa \end{aligned}\right\} \tag{6}$$

for t and x. Then

$$\left.\begin{aligned} \frac{\partial\Pi}{\partial t} &= \frac{\partial\Pi}{\partial s}\frac{\partial s}{\partial t}+\frac{\partial\Pi}{\partial z}\frac{\partial z}{\partial t} = -\kappa(x-1)\,e^{-\kappa t}\frac{\partial\Pi}{\partial s} \\ \text{and}\qquad \frac{\partial\Pi}{\partial x} &= \frac{\partial\Pi}{\partial s}\frac{\partial s}{\partial x}+\frac{\partial\Pi}{\partial z}\frac{\partial z}{\partial x} = e^{-\kappa t}\frac{\partial\Pi}{\partial s}+\frac{\lambda}{\kappa}\frac{\partial\Pi}{\partial z}, \end{aligned}\right\} \tag{7}$$

so that (5) becomes simply $\Pi = \partial\Pi/\partial z$ of which the solution is obviously $\Pi(s, z) = e^z\Phi(s)$, where Φ is an arbitrary function. Consequently, in the original arguments, t and x,

$$\Pi(x, t) = e^{\lambda x/\kappa}\,\Phi\{(x-1)\,e^{-\kappa t}\}. \tag{8}$$

The form of Φ must be determined by the boundary conditions at $t = 0$, when we know there were N particles, so that $p_N(0) = 1$ and $\Pi(x, 0) = x^N$. Substituting $t = 0$ and this result in (8) gives

$$x^N = e^{\lambda x/\kappa}\,\Phi(x-1),$$

hence, with $y = x-1$, $\Phi(y) = (1+y)^N e^{-\lambda(1+y)/\kappa}$ and finally

$$\Pi(x, t) = \{1+(x-1)\,e^{-\kappa t}\}^N \exp\left\{\frac{\lambda}{\kappa}(x-1)\,(1-e^{-\kappa t})\right\}. \tag{9}$$

It is convenient to denote the number of particles in the region at time t by $n(t)$, so that $n(0) = N$; and the result just proved can be expressed as

Theorem 2. *In an immigration–emigration process the distribution of $n(t)$, given $n(0) = N$, has probability generating function given by* (9).

Since a probability generating function characterizes a distribution, this result solves the problem of determining the distribution of the number at any time instant, and any features of the distribution such as means and variances can be found (see below). The joint distribution at several time points is also discussed below.

Consider next the behaviour of (9) as $t \to \infty$, that is a long time from the start. Since $\kappa > 0$, $\lim_{t\to\infty} \Pi(x, t) = \Pi(x)$, say, exists and

$$\Pi(x) = \exp\left\{\frac{\lambda}{\kappa}(x-1)\right\}, \tag{10}$$

which is (equation 2.6.11), the probability generating function of a Poisson distribution of parameter λ/κ, $P(\lambda/\kappa)$. The same distribution can also be produced as a non-limiting result. Suppose $n(0)$ has the distribution corresponding to (10), that is $P(\lambda/\kappa)$. Then (9) is the probability generating function of $n(t)$ for fixed $n(0) = N$ and theorem 3.1.1 shows that the unconditional probability generating function of $n(t)$ is

$$\sum_N \{1+(x-1)e^{-\kappa t}\}^N \exp\left\{\frac{\lambda}{\kappa}(x-1)(1-e^{-\kappa t})\right\} e^{-\lambda/\kappa}(\lambda/\kappa)^N/N!$$

$$= \exp\left\{\frac{\lambda}{\kappa}(x-1)(1-e^{-\kappa t})-\frac{\lambda}{\kappa}\right\}\sum_N \left\{\frac{\lambda}{\kappa}+\frac{\lambda}{\kappa}(x-1)e^{-\kappa t}\right\}^N\Big/N!$$

$$= \exp\left\{\frac{\lambda}{\kappa}(x-1)\right\},$$

which is (10) again. We therefore have

Theorem 3. *In an immigration–emigration process the probability distribution of $n(t)$ tends, as $t \to \infty$, to $P(\lambda/\kappa)$, irrespective of the initial conditions. If the distribution of $n(0)$ is $P(\lambda/\kappa)$ then so is the unconditional distribution of $n(t)$ for all t.*

For any stochastic process, $\{n(t)\}$, a distribution of $n(t)$ which does not change with t is called a *stationary*, or *equilibrium* distribution. Here $P(\lambda/\kappa)$ is such a distribution. It can be found directly, assuming it to exist, without calculating the conditional distribution (9), by allowing t to tend to infinity in (2). If

$$\lim_{t\to\infty} p_n(t) = p_n$$

exists it must satisfy

$$(\lambda+n\kappa)p_n = \lambda p_{n-1}+\kappa(n+1)p_{n+1}, \tag{11}$$

which may be solved in a variety of ways: perhaps the best is by generating functions again, which leads to (5) with $\partial\Pi/\partial t = 0$, that is to

$$\lambda\Pi = \kappa d\Pi/dx \tag{12}$$

with (10) as the obvious solution with boundary condition $\Pi(1) = 1$.

The exponential distribution

Theorem 1 has many uses besides the specific one made here. Consider a process of incidents, that is a process with the same sample space as a Poisson process (§2.3), with probabilities defined as follows: the intervals between successive incidents are independent random variables with a common exponential distribution of parameter κ. Then theorem 1 says that the probability of an incident in any interval $(t, t+\delta t)$ is $\kappa\delta t + o(\delta t)$, independent of how long it was since the last incident occurred. Furthermore, since the intervals between earlier incidents are also independent of the interval between incidents in which t occurs, the probability of an incident in $(t, t+\delta t)$ is independent of whatever incidents took place prior to t. Hence the process so described is a Poisson process. Theorem 1 therefore provides another means of defining a Poisson process and is a sort of converse to theorem 2.3.2.

Examples of immigration–emigration processes

The immigration–emigration process was first studied in connexion with telephone engineering by Erlang (see Brockmeyer *et alii* 1948). The necessary change of language is that the region is a telephone exchange, the particles are subscribers and staying in the region means making a call. The model assumes that the demands for service form a Poisson process, which is reasonable, that the durations of calls have an exponential distribution, which is often unreasonable, and that the exchange can cope with any number of calls. The final assumption is clearly not true and the effect of delays due to the exchange being saturated will be considered under queueing processes (§§4.2, 4.3). Other applications are: (i) The motion of particles in a region, in the language we have used, though many particles move in a way that contradicts the exponential stay assumption. (ii) The size of a population of animals, where the possible changes are arrivals of new animals (immigration) and their departure (by emigration or death). A further factor may be births, causing new arrivals, but these would presumably increase in rate the larger $n(t)$ was, whereas the arrivals in our model come at a constant rate. The process can be generalized to a 'birth–death–immigration' process by adding extra terms to (1). (iii) The thickness and strength of yarns. In this application the former time variable is spatial, being the distance along the thread, the arrival is the left-hand end-point of a fibre, the departure is the right-hand end-point. The number of particles in a region is the number of fibres in a cross-section of the thread. The fibres occur at random along the thread length and have an exponential distribution of length: both quite reasonable assumptions for some fibres, such as jute.

Use of the generalized addition law

There are some details in the derivation of (1) that are worth attention. The generalized addition law (theorem 1.4.4) is used with all probabilities conditional on the event $n(0) = N$: in full it is

$$\begin{aligned} p[n(t) = n \mid n(0) = N] &\\ = \sum_{s=0}^{\infty} p[n(t) = n \mid n(t-\delta t) = s \quad \text{and} \quad n(0) = N]&\\ \times p[n(t-\delta t) = s \mid n(0) = N], \quad (13)\end{aligned}$$

since the events '$n(t-\delta t) = s$' for different s are exclusive and exhaustive. Now, because of theorem 1, and the independence of the different particles, the behaviour of the process in the interval $(t-\delta t, t)$ depends only on $n(t-\delta t)$, the number present at the beginning of the interval. Hence

$$\begin{aligned} p[n(t) = n \mid n(t-\delta t) = s \quad \text{and} \quad n(0) = N]&\\ = p[n(t) = n \mid n(t-\delta t) = s], \qquad (14)\end{aligned}$$

and all these probabilities are $o(\delta t)$ except when $s = n-1$, n or $n+1$. When $s = n-1$, the transition can take place by one particle arriving and none leaving. The arrival has probability $\lambda\delta t$. The probability of no departure for any one particle is $e^{-\kappa\delta t} = 1-\kappa\delta t$ to this order, and hence for $(n-1)$ particles

$(1-\kappa\,\delta t)^{n-1} = 1-\kappa(n-1)\,\delta t$ to this order. These probabilities are independent so this transition has probability

$$\lambda\,\delta t\{1-\kappa(n-1)\,\delta t\} = \lambda\,\delta t,$$

to this order. Notice the way in which one event of probability of order δt (here $\lambda\,\delta t$) effectively means that the other events (departures) need not be considered. The transition can also take place by two arriving and one leaving but, as is easily seen, this is $o(\delta t)$. When $s = n+1$ the only transition that matters is one departure and no arrivals. A departure has probability $(n+1)\,\kappa\,\delta t$, by the independence of the particles, and hence the arrival probability (in fact $1-\lambda\,\delta t$) need not be considered. With $s = n$ the major probability is of no arrivals or departures, namely $(1-\lambda\,\delta t)\,(1-n\kappa\,\delta t) = 1-\lambda\,\delta t-n\kappa\,\delta t$. Hence (1) follows to order δt. The passage from (1) to (2) is carried out by collecting together terms of like order and then passing to the limit as $\delta t \to 0$. The resulting differential-difference equation is then turned into a partial difference equation by the use of generating functions. These procedures are of wide applicability.

Use of the probability generating function

From (9) it is possible to calculate any desired feature of the distribution: for example, expansion in a power series in x will give the probabilities $p(n(t) = n\,|\,n(0) = N)$. If $x = e^z$ a similar expansion in terms of z yields the moments about the origin (equation 2.6.4). The same expansion of $\ln\Pi(e^z, t)$ gives the cumulants (equation 2.6.7). For example, let us evaluate the coefficient of z in the last expansion: that is, κ_1, the mean, obtained by differentiating $\ln\Pi(e^z, t)$ once with respect to z and putting $z = 0$. Now

$$\ln\Pi(e^z, t) = N\ln\{1+(e^z-1)\,e^{-\kappa t}\}+(\lambda/\kappa)\,(e^z-1)\,(1-e^{-\kappa t}),$$

so
$$\{\partial\ln\Pi(e^z, t)/\partial z\}_{z=0} = Ne^{-\kappa t}+(\lambda/\kappa)\,(1-e^{-\kappa t}), \qquad (15)$$

or
$$\mathscr{E}\{n(t)\,|\,n(0)\} = (\lambda/\kappa)\,(1-e^{-\kappa t})+e^{-\kappa t}\,n(0), \qquad (16)$$

and the regression of $n(t)$ on $n(0)$ is linear with regression coefficient $e^{-\kappa t}$ (§3.1). A slight rewriting of (16) as

$$e^{-\kappa t}\{n(0)-\lambda/\kappa\}+\lambda/\kappa$$

shows that if $n(0) > \lambda/\kappa$: that is, if the number at the beginning exceeds the expected number as $t \to \infty$ (the mean of the limiting Poisson distribution), then the expected number at t exceeds λ/κ. In words: an above-average number at the beginning leads one to expect an excess above average at any later time. The effect diminishes sharply with time due to the exponential factor $e^{-\kappa t}$.

Limiting distribution

The limiting distribution as $t \to \infty$ is important. It means that if the process is considered a long way from the start the distribution of $n(t)$ will be of the simple Poisson form instead of the complicated form in (9) and, moreover, will not depend on the conditions at the start. It is not the *number* in the region that tends to a limit, indeed it fluctuates up and down, but the *distribution* of the number. The proof requires the use of theorem 3.6.4 to show that the limit of the distribution is implied by the limit of the generating function: we have only to put $x = e^{it}$ in (9) to apply that result.

It is natural to ask what happens if the process starts in the limiting situation. In analogy with limits of iterative sequences one would expect it to stay there; for example, in the usual numerical process for finding the square root of a number c, where a sequence $\{a_n\}$ is defined by $a_n = \frac{1}{2}(a_{n-1} + c/a_{n-1})$, with an arbitrary start a_0, and $\lim_{n\to\infty} a_n = \sqrt{c}$; $\therefore a_0 = \sqrt{c}$ then $a_n = \sqrt{c}$ for all n. The calculations show that this is true here, if the process starts with $P(\lambda/\kappa)$ it remains at this distribution. This means that in processes with this property it is often possible to ignore any 'start', because it is the same as any other point of time, and think of the processes for all time. The distribution at any time will be Poisson, and the conditional distribution at any two points of time separated by an interval of length t will be given by (9). Such processes are called stationary and will be formally defined below. Notice that the stationarity refers to the *distributions* of the process at different time points, and not to the values of the process itself. In fig. 4.1.1 (*a*–*c*) we show graphs of $n(t)$ against t obtained in samples from processes with different values of λ and κ. They were obtained by taking the intervals between arrivals to be random samples from an exponential distribution with parameter λ, and the times of stay of each particle to be the same with parameter κ, in the manner

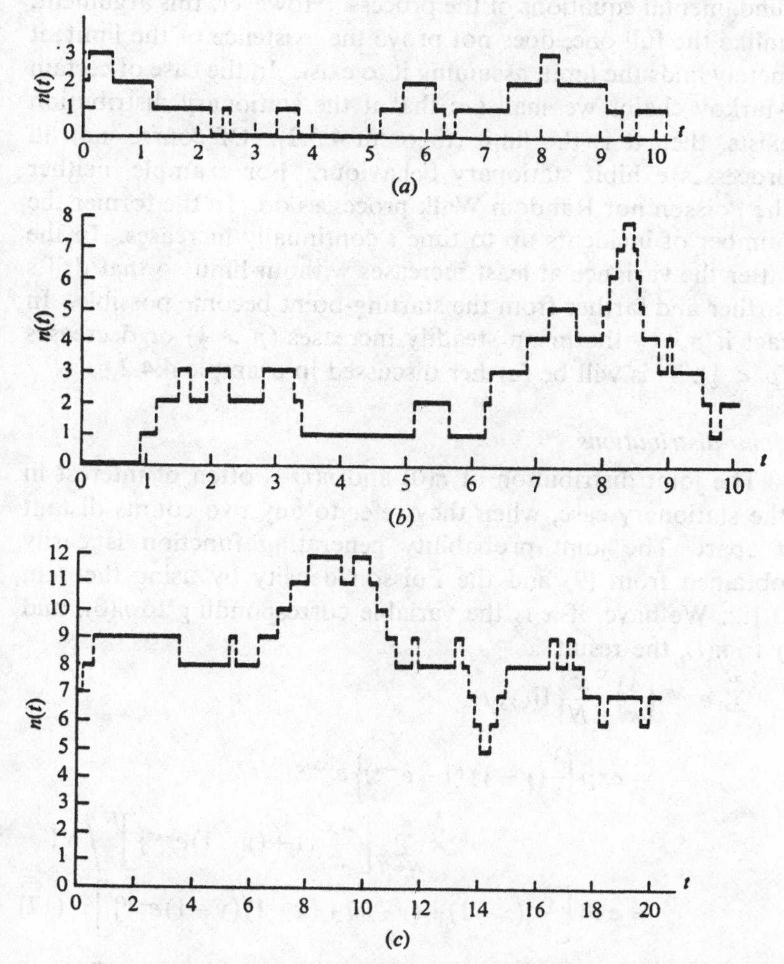

Fig. 4.1.1 (*a–c*). Realizations of immigration–emigration processes.

described in §3.5. With the arrivals and departures so recorded and a random sample from $P(\lambda/\kappa)$ to start the process, the numbers in the region at each point of time can be found. Such graphs are called *sample paths* or *realizations* of the process.

Often it is enough to consider the stationary behaviour of a process and then the distributions may be obtained directly, as in (11) above, by ignoring the dependence on time in the fundamental equations of the process. However, this argument, unlike the full one, does not prove the existence of the limit; it merely finds the limit, assuming it to exist. In the case of certain Markov chains we shall see that if the stationary distribution exists, then it is the limit (theorem 4.5.2). Of course, not all processes exhibit stationary behaviour. For example, neither the Poisson nor Random Walk processes do. In the former the number of incidents up to time t continually increases. In the latter the variance at least increases without limit so that drifts farther and farther from the starting-point become possible. In fact if $p \neq \frac{1}{2}$ the mean steadily increases ($p > \frac{1}{2}$) or decreases ($p < \frac{1}{2}$). This will be further discussed in example 4.4.2.

Joint distributions

The joint distribution of $n(0)$ and $n(t)$ is often of interest in the stationary case, when they refer to any two counts distant t apart. The joint probability generating function is easily obtained from (9) and the Poisson density by using theorem 3.1.2. We have, if x is the variable corresponding to $n(0)$, and y to $n(t)$, the result

$$\begin{aligned}
\sum_{N=0}^{\infty} e^{-\lambda/\kappa}\left(\frac{\lambda}{\kappa}\right)^N \frac{x^N}{N!}\Pi(y,t) &= \exp\left\{\frac{\lambda}{\kappa}(y-1)(1-e^{-\kappa t})\right\} e^{-\lambda/\kappa} \\
&\quad \times \sum_{N=0}^{\infty}\left[\frac{\lambda x}{\kappa}\{1+(y-1)e^{-\kappa t}\}\right]^N \Big/ N! \\
&= \exp\left[\frac{\lambda}{\kappa}\{(x-1)+(y-1)+(x-1)(y-1)e^{-\kappa t}\}\right]. \quad (17)
\end{aligned}$$

This is the probability generating function of a *bivariate Poisson distribution*. From (17) probabilities and moments may be calculated as with (9) but the results are not usually simple. The covariance of $n(0)$ and $n(t)$ may easily be found by exploiting the fact that the regression of $n(t)$ on $n(0)$ is linear with regression coefficient $e^{-\kappa t}$, (16). Then the covariance is (equation 3.1.21)

$e^{-\kappa t}(\lambda/\kappa)$ since the variance of $n(0)$ is λ/κ. The correlation between $n(0)$ and $n(t)$ is $e^{-\kappa t}$, and is called the *autocorrelation function*, $\rho(t)$, of the process. Notice that at $t = 0$ it is necessarily 1 and as t increases it diminishes exponentially fast. By use of a bivariate form of the limit theorem 3.6.4 it is possible to show that as λ/κ, the expected number of particles in the region, tends to infinity, the bivariate Poisson distribution tends to the bivariate normal distribution. (In §3.6 we remarked on the corresponding univariate result that the ordinary $P(\mu)$ distribution tended to normality as $\mu \to \infty$.) Consequently, if the expected number of particles in a region is large we can say that the joint distribution of $n(0)$ and $n(t)$ is approximately bivariate normal with means λ/κ and dispersion matrix

$$\begin{pmatrix} \lambda/\kappa & \lambda e^{-\kappa t}/\kappa \\ \lambda e^{-\kappa t}/\kappa & \lambda/\kappa \end{pmatrix}, \tag{18}$$

and consequently the autocorrelation $\rho(t) = e^{-\kappa t}$ is a good guide to the association between $n(0)$ and $n(t)$. Occasionally the joint distribution of $n(t_1), n(t_2), \ldots, n(t_k)$ in the stationary case is required. This is easily obtained and we illustrate with $k = 3$. We have, with $t_1 < t_2 < t_3$, for the joint density of $n(t_i)$, in an obvious, abbreviated notation,

$$\begin{aligned} p(n_1, n_2, n_3) &= p(n_3 \mid n_1, n_2)\, p(n_1, n_2) \\ &= p(n_3 \mid n_1, n_2)\, p(n_2 \mid n_1)\, p(n_1) \\ &= p(n_3 \mid n_2)\, p(n_2 \mid n_1)\, p(n_1). \end{aligned} \tag{19}$$

The final line follows since the development of the process in the interval (t_2, t_3) depends only on $n(t_2) = n_2$ and is independent of $n(t_1) = n_1$. This is the Markov property to be discussed in §4.5. Since the two conditional distributions in (19) are known from (9)—for example, $p(n_3 \mid n_2)$ is obtained with $N = n_2$ and $t = t_3 - t_2$ because, due to the stationarity, the origin of time does not matter—and $p(n_1)$ is Poisson, the joint distribution of the three variables is known. A practically useful approximation for large λ/κ is that $n(t_1), n(t_2), \ldots, n(t_k)$ are multivariate normal with means λ/κ, variances λ/κ and the correlation between $n(t_i)$ and $n(t_j)$ equal to

$$e^{-\kappa|t_j - t_i|} = \rho(|t_j - t_i|).$$

Stationarity

We can now give a formal definition of stationarity. Suppose that each point of sample space has associated with it a function $x(t)$ defined for each t belonging to some set T. That is, each elementary event determines a realization, as described above. Suppose, further, that for any finite number, n, of values of t, $t_1 < t_2 < \ldots < t_n$ the joint distribution of $x(t_1), x(t_2), \ldots, x(t_n)$ is defined: thus they form a set of random variables. $x(.)$ is then said to be a stochastic process, and it is *temporally homogeneous* if the joint distribution just described, conditional on $x(t_1)$, depends only on $t_2 - t_1, t_3 - t_2, \ldots, t_n - t_{n-1}$ and not on t_1. A temporally homogeneous process is *stationary* if the distribution of $x(t)$ does not depend on t. T is nearly always either the real line, as in this section, or the set of integers, as with the random walk. In the latter case we usually write x_n instead of $x(n)$. All the processes studied in this book are temporally homogeneous. This property means that given the value of the process at any point of time, t_1, then t_1 may be taken as the origin: the future development of the process will not be affected. This property clearly obtains for the random walk: given that it is at the origin, for example, it does not matter when in time it is there, the future development is the same. This would not be true, if, for example, the probability of moving to the right at the nth step depended on n. Stationarity demands, in addition, that the distribution is the same at all points of time. This is not true of the random walk, but is true of the immigration–emigration process provided the distribution at any one point of time is $P(\lambda/\kappa)$. For a stationary process the origin of time is completely irrelevant. Many stationary processes exhibit an ergodic property (§3.6). That is, their behaviour may be studied, either by considering several realizations at a fixed point of time, or by considering a single realization over many points of time, with equivalent results. The property will be used in later sections of this chapter. The term stochastic process is also used for the whole sample space and not merely for certain random variables associated with it. For example, in the next section we study a queueing process for which several random variables, $x(t)$, are defined: thus, the queue size and the waiting-times.

4.2. Simple queueing process

Consider a Poisson process of rate λ in which the incidents are arrivals of 'customers' at a 'server'. If, when a customer arrives, the server is 'free' then he immediately serves the customer for a period s called the customer's service-time. If, on the contrary, the server is 'busy' with another customer, the customers join a queue in order of their times of arrival and as soon as the server finishes with one customer he attends to the customer at the head of the queue. This 'discipline' in the behaviour of the queue is called 'first come, first served'. The service-times are independent random variables with a common exponential distribution with parameter κ, and are independent of the Poisson arrival process. This is a *simple queueing process* with a single server.

Denote by $p_n(t)$ the probability that at time t there are n customers in the queue (including the one being served†). This is conditional on the initial conditions at $t = 0$, the start, but we shall only find the limiting stationary distribution as $t \to \infty$ for which, as explained in the previous section, these are not relevant. Then the probability that a customer will join the queue in an interval of length δt is $\lambda \delta t + o(\delta t)$, because of the Poisson process of arrivals; the probability that a customer will leave the queue is $\kappa \delta t + o(\delta t)$, if $n > 0$, and is otherwise zero, because of the exponential distribution of service-time (theorem 4.1.1): and these probabilities are independent of any events that concern only the process before the commencement of the small interval. Consequently we can write down, using the same techniques as were used to obtain equation 4.1.1, for $n > 0$,

$$p_n(t) = p_{n-1}(t-\delta t)\,\lambda\delta t + p_{n+1}(t-\delta t)\,\kappa\delta t + p_n(t-\delta t)\,(1-\lambda\delta t-\kappa\delta t) \quad (1)$$

and, for $n = 0$,

$$p_0(t) = p_1(t-\delta t)\,\kappa\delta t + p_0(t-\delta t)\,(1-\lambda\delta t). \quad (2)$$

† This convention that the queue includes the customer being served will be adhered to.

These give, on passage to the limit, $\delta t \to 0$,

$$dp_n(t)/dt = \lambda p_{n-1}(t) + \kappa p_{n+1}(t) - (\lambda + \kappa)\, p_n(t), \tag{3}$$

for $n > 0$, and

$$dp_0(t)/dt = \kappa p_1(t) - \lambda p_0(t). \tag{4}$$

Assuming $\lim_{t\to\infty} p_n(t) = p_n$ to exist, the left-hand sides of the equations can be put equal to zero and the argument t omitted from the right. Addition of the equations for $n = 0, 1, \ldots, r-1$ gives $\kappa p_r = \lambda p_{r-1}$, or $p_r = (\lambda/\kappa)\, p_{r-1}$ and

$$p_r = (\lambda/\kappa)^r\, p_0. \tag{5}$$

This can only be a probability distribution if $\rho = \lambda/\kappa < 1$, when p_0 must be $(1-\rho)$, in order that $\Sigma p_r = 1$. Hence a limit can only exist if $\rho < 1$. It is possible to show that if $\rho < 1$ the limit does exist and hence we have

Theorem 1. *A simple queueing process with $\rho < 1$ has a stationary distribution of queue size given by* (5) *with $p_0 = 1-\rho$.*

This distribution is the geometric distribution with parameter ρ, which we write $G(\rho)$ (equation 2.1.10). ρ is called the *traffic intensity*. Since the mean of an exponential distribution of parameter κ is κ^{-1} (theorem 2.3.6 with $n = 1$) the traffic intensity is the ratio of the mean service time, κ^{-1}, to the mean of the intervals between arrivals, λ^{-1}. The mean of $G(\rho)$ is $\rho/(1-\rho)$ (equation 2.1.11), the average size of queue, including the customer being served. If he is excluded the average size is clearly

$$\sum_{r=1}^{\infty} (r-1)p_r = (1-\rho)\rho^2\Sigma(r-1)\rho^{r-2}$$

$$= \rho^2/(1-\rho). \tag{6}$$

The probability that the server is free is $p_0 = 1-\rho$.

Notice that p_r is the probability that r people are in the queue in the stationary distribution. Consider instead the probability that when a customer arrives he will find r people in the queue: that is $p(\text{queue size } r \text{ at } t \mid \text{arrival in } (t, t+\delta t))$. It is not immediately obvious that this is still p_r, the unconditional probability. To see that it is, we need only remark that the event, A, of an arrival is independent of the event, B, that the queue size is r, because the latter depends on arrivals prior to t, and A is independent of these arrivals by the definition of a Poisson process. The independence of events A and B means

$$p(B \mid A) = p(B)$$

and hence the conditional probability is p_r. With this remark it is easy to find the distribution of queueing-time of a customer: that is, the time q between arrival and completion of service. Let $f(q)$ be the density of q and $f(q \mid r)$ the density of q conditional on r customers being in the queue when the customer arrives. Then by the remark and a discrete variable† form of equation 3.2.7

$$f(q) = \sum_{r=0}^{\infty} f(q \mid r) p_r. \tag{7}$$

But $f(q \mid r)$ is easily found: the customer has to queue during the remaining service-time of the customer being served when he arrives and during r other complete service-times (including his own). By theorem 4.1.1 the remaining service-time still has an exponential distribution of parameter κ. By theorem 2.3.5 the sum of $(r+1)$ independent service-times has a $\Gamma(r+1, \kappa)$ distribution, so $f(q \mid r)$ has that density. Finally, inserting the explicit form for the Γ-density,

$$\begin{aligned} f(q) &= \sum_{r=0}^{\infty} e^{-\kappa q} \kappa^{r+1} q^r \rho^r (1-\rho)/r! \\ &= \kappa(1-\rho) e^{-\kappa(1-\rho) q}, \end{aligned} \tag{8}$$

an exponential distribution of parameter $\kappa(1-\rho)$. Hence

Theorem 2. *The distribution of queueing-time in a simple queueing process is exponential with parameter* $\kappa(1-\rho)$.

Examples of simple queueing processes

The simple queueing process was also first studied by Erlang in telephone engineering. In the language appropriate to that application, the difference between this process and the immigration–emigration process of the last section is that, in the

† We are here discussing certain properties of the joint distribution of a continuous random variable, q, and a discrete one, r. Compare the proof of theorem 2.3.3.

latter, requests for a connexion were supposed always met immediately, whereas in this process there is supposed only one line at the exchange and the callers wait in order of arrival for their connexion. The extension to the case of $n > 1$ lines at the exchange presents no difficulties beyond complexity of algebra. But queues occur in many other situations: aircraft waiting to land at an airport, ships waiting to enter a port, machines waiting to be repaired, patients waiting to see a doctor, and many others. The simple queueing process discussed here makes assumptions that may well not be satisfied in some applications. For example, aircraft run to schedules and would not arrive in a Poisson process; the theory based on the exponential service-time distribution is unlikely to be useful when unloading ships. Nevertheless, the theory has found a wide range of applications.

Existence of a stationary distribution

The method of deriving the basic equations (1) and (2), is the same as before (§4.1). Their general solution is complicated so we pass instead directly to the stationary case where the equations are almost trivial. Contrary to the immigration–emigration process these equations do not always have a *probability* solution, that is a solution with $p_r \geqslant 0$ and $\Sigma p_r = 1$. The geometric series only converges when $\lambda/\kappa < 1$. Then the solution is as stated. We do not prove that the limit is attained, nor do we investigate the case of $\rho \geqslant 1$. The practical situation with $\rho \geqslant 1$ is that very large queues build up and $p_n(t) \to 0$ as $t \to \infty$ for each n. Thus the value $\rho = 1$ is critical, the behaviour on one side of it being quite different from the other side. In the language to be developed in §§4.4 and 4.5 in connexion with slightly different processes the state 'queue size n' is transient if $\rho > 1$, null if $\rho = 1$, and ergodic if $\rho < 1$. The interpretation of ρ, the traffic intensity (usually measured in units called erlangs), given above as

$$\frac{\text{mean service-time}}{\text{mean inter-arrival time}}$$

shows why $\rho = 1$ is critical. If $\rho > 1$ then in an average service-time (the time which one customer takes to be attended to) more than one customer will arrive: the server cannot cope with the demands and large queues will build up. On the other hand, if $\rho < 1$ fewer than one customer will arrive and the server will have some leisure, known as *slack periods*. The probability of his being slack is $1-\rho$, when the queue size is zero, the most probable size of queue. The expected queue size is $\rho/(1-\rho)$ which increases sharply as $\rho \to 1$. Hospital authorities appear to work with ρ very near 1 in order that the doctor never be kept waiting ($p_0 = 1-\rho$). This causes large queues since the mean is $\rho/(1-\rho)$. Any reasonable system should keep ρ away from 1 even though the server will have some slack periods.

Queueing-time

From the customer's point of view the important thing is the queueing-time. How long is it going to take him to get through the system? The answer, equation (8), is encouraging in that the shorter times are the more frequent because of the exponential distribution. The expected queueing-time is $\{\kappa(1-\rho)\}^{-1}$ also increasing sharply as ρ approaches 1. The *waiting-time*, w, defined as the time between arrival and being served, is sometime more important—women choosing a dress do not mind a long service time but they do not like to be kept waiting—and is interesting because the distribution function is a mixture of discrete and continuous parts (§2.2). The probability of finding the server free on arrival is, by the remark above, $p_0 = 1-\rho$. So the probability that $w = 0$ is $1-\rho$. If the server is not free and r (> 0) customers are in the queue, then the customer will have to wait through r service-times and by the same argument that produced (8) the distribution will have density for $w > 0$

$$\kappa\rho(1-\rho)e^{-\kappa(1-\rho)w}. \tag{9}$$

The distribution function of w is therefore

$$\left.\begin{aligned} F(w) &= 0 \quad (w < 0);\\ F(w) &= (1-\rho)+\rho(1-e^{-\kappa(1-\rho)w}) \quad (w \geqslant 0); \end{aligned}\right\} \tag{10}$$

with a discontinuity, equal to $(1-\rho)$, at $w = 0$. The expected waiting-time is $\rho/\kappa(1-\rho)$.

Ergodicity

The probability that the server is free (in the stationary distribution) is $p_0 = 1-\rho$. This means that if the process is started at $t = 0$ with any number of customers present, then, a long time, t, after the start, the probability that the server will be free is $1-\rho$; or, in practical terms, if the process is started up a large number of times the frequency of times the server will be found free at t is about $1-\rho$. But the probability has another practical interpretation in terms of a *single* realization; namely, over a long time interval of length t the amount of time that the server will be free will be about $(1-\rho)t$. We shall not prove this result, but it is of considerable importance and is another example of an *ergodic* theorem (§3.6). The point is discussed in connexion with recurrent events in §4.4. The first interpretation is a spatial average and the second a temporal average. Similar temporal interpretations are available for the other probabilities.

Busy periods

The ergodic principle can often be used to find probabilities or expectations. The slack periods of the server have already been mentioned: these alternate in time with *busy periods* during which the server is continuously occupied. Now the probability distribution of the length of slack period is clearly $E(\lambda)$, because of the Poisson process of arrivals; and in particular the expected length of a slack period is λ^{-1}. Now consider a single realization with its alternation of busy and slack periods of lengths $b_1, s_1, b_2, s_2, \ldots$. By the strong law of large numbers† (theorem 3.6.3—an ergodic result),

$$\lim_{n\to\infty}\left(\sum_{i=1}^{n} s_i \Big/ n\right) = \lambda^{-1}.$$

By the ergodic result just mentioned,

$$\lim_{n\to\infty}\left(\sum_{i=1}^{n} s_i \Big/ \sum_{i=1}^{n} b_i\right) = p_0/(1-p_0) = (1-\rho)/\rho.$$

Hence $$\lim_{n\to\infty}\left(\sum_{i=1}^{n} b_i \Big/ n\right) = \rho/[(1-\rho)\,\lambda] = (\kappa-\lambda)^{-1}$$

† All the results in the rest of this paragraph require the qualification 'with probability one'.

and by a final ergodic result, the converse of the strong law, $\mathscr{E}(b_i) = (\kappa - \lambda)^{-1}$, the expected length of a busy period. Notice that it tends to infinity as $\rho \to 1$ for fixed κ. More elaborate methods are needed to find the density of the busy period distribution. The expected number served in a busy period is discussed in §4.3.

4.3. Queueing process with Poisson input and general service distribution

Consider a stochastic process defined as in the last section except that the service times are independent random variables with a common distribution, independent of the Poisson arrival process. The simple queueing process is the special case where the distribution of service-times is $E(\kappa)$.

Consider the probability that a departing customer leaves behind him a queue of j customers, conditional on the event that the previous customer left behind him a queue of i customers: here i and j are non-negative and since the change from i to j depends only on the Poisson arrivals in between the two customers' departures, the probability will not depend on the particular customer (nor on t) and may simply be denoted by p_{ij}. We derive an expression for p_{ij} in terms of known quantities. Consider first p_{0j}. The customer leaving j behind him must have arrived and found the server free, since the previous customer left him so, and hence these j must have arrived during this customer's service-time. Consequently

$$p_{0j} = p(j \text{ customers arrive during one customer's service-time})$$
$$= \int_0^\infty p(j \text{ customers arrive} \mid \text{service-time is } x) f(x)\,dx, \quad (1)$$

where $f(x)$ is the given density of service-times. (If the distribution is discrete a summation, $\sum_r$, replaces the integration in (1), and p_r replaces $f(x)$.) This is again a consequence of the generalized addition law in its integral form (equation 3.2.7). The probability of j given x is $P(\lambda x)$ (theorem 2.3.3), so that

$$p_{0j} = \int_0^\infty e^{-\lambda x}(\lambda x)^j f(x)\,dx/j! = \pi_j, \text{ say.} \quad (2)$$

Similarly, when $i \neq 0$, $p_{ij} = p(j-i+1$ customers arrive during one customer's service-time) since when $i\,(>0)$ customers are left behind $(i-1)$ are still there when the next customer leaves, so $j-(i-1)$ must have arrived to bring the number up to j. Hence, if $i \neq 0$,

$$p_{ij} = \pi_{j-i+1} \quad \text{if} \quad j > i-1; \quad \text{otherwise } p_{ij} = 0. \tag{3}$$

Now let $p_n^{(s)}$ be the probability that the customer number s leaves a queue of n behind him when he leaves; the customers being supposed numbered in order of their arrival and this probability, unlike p_{ij}, being unconditional. The process may be supposed to start with no one in the queue, but since we shall only find the stationary distribution, the initial conditions do not matter. To find the equilibrium value of $p_n^{(s)}$, $p_n = \lim_{s\to\infty} p_n^{(s)}$, we proceed using the same method as in §§4.1 and 4.2, except that instead of relating the state of the process now, at time t, with the state in the past, at time $t-\delta t$, we relate the state now, as the sth customer leaves, with the state in the past when the $(s-1)$st customer left. By the original form of the generalized addition law

$$p_j^{(s)} = \sum_{i=0}^{\infty} p_i^{(s-1)} p_{ij} \quad (j \geq 0), \tag{4}$$

and hence in equilibrium, where the dependence on s is omitted,

$$p_j = \sum_{i=0}^{\infty} p_i p_{ij} = (p_0+p_1)\pi_j + p_2\pi_{j-1} + \dots + p_{j+1}\pi_0. \tag{5}$$

This difference equation is more complicated than the trivial one for the simple queue and is most easily solved using probability generating functions. Introduce two of these, noting that the π's form a probability distribution,

$$P(x) = \sum_0^{\infty} p_j x^j \quad \text{and} \quad \Pi(x) = \sum_0^{\infty} \pi_j x^j, \tag{6}$$

multiply equation (5) by x^j and sum over j from zero to infinity. The result is

$$\begin{aligned} P(x) &= \Pi(x)\,[p_0+p_1+p_2x+p_3x^2+\dots] \\ &= [p_0(x-1)+P(x)]\,\Pi(x)/x, \end{aligned}$$

whence

$$P(x) = \frac{\Pi(x)\,(x-1)}{x-\Pi(x)}\,p_0. \tag{7}$$

Now $\Pi(x)$ is known; in fact from (2)

$$\Pi(x) = \sum_{j=0}^{\infty} \int_0^\infty e^{-\lambda y}(\lambda y)^j f(y)\,dy\,.\,x^j/j!$$

$$= \int_0^\infty e^{-\lambda y} f(y)\,[\sum_j (\lambda x y)^j/j!]\,dy$$

$$= \int_0^\infty e^{\lambda y(x-1)} f(y)\,dy = \Phi[\lambda(x-1)], \tag{8}$$

where $\Phi(z)$ is the *moment* generating function (equation 2.6.10) of the service-time distribution. Consequently $P(x)$ is known apart from the value of p_0. But $P(x)$ must be a probability generating function and so $\lim_{x\to 1} P(x) = 1$. The limit of the right-hand side of (7) can be found by differentiating the numerator and denominator and then considering the limit of the ratio of these: this is the same as the original limit. Carrying this out we obtain

$$\lim_{x\to 1} \frac{\Pi'(x)\,(x-1)+\Pi(x)}{1-\Pi'(x)}\,p_0,$$

and from (8) $\Pi(1) = \Phi(0) = 1$ and $\Pi'(1) = \lambda\Phi'(0) = \lambda/\kappa$ where, as in §4.2, κ^{-1} is the mean of the service-time distribution (see equation 2.6.4). Consequently, returning to (7), with $\rho = \lambda/\kappa$

$$1 = \frac{1}{1-\rho}\,p_0$$

and

$$p_0 = (1-\rho). \tag{9}$$

For this to be possible we must have $\rho < 1$. Hence we have

Theorem 1. *In a queueing process with Poisson input of rate λ, and independent service times having a common distribution with moment generating function $\Phi(z)$, mean κ^{-1} and variance σ^2, the equilibrium distribution of the size of queue left by a departing customer has, if $\rho = \lambda/\kappa < 1$, probability generating function*

$$\sum_0^\infty p_n x^n = P(x) = \frac{(x-1)\,\Phi[\lambda(x-1)]}{x-\Phi[\lambda(x-1)]}\,(1-\rho). \tag{10}$$

In particular $p_0 = 1-\rho$, and the expected queue size is

$$\{\rho(2-\rho)+\lambda^2\sigma^2\}/2(1-\rho). \tag{11}$$

Equation (11) follows on evaluating $P'(1)$: this is most easily done by writing (7) in the form

$$P(x)\{x-\Pi(x)\} = \Pi(x)(x-1)p_0,$$

differentiating twice and putting $x = 1$. From (8)

$$\Pi''(1) = \lambda^2\Phi''(0) = \lambda^2\mathscr{E}(s^2)$$

(equation 2.6.4), and $\mathscr{E}(s^2) = \sigma^2+\kappa^{-2}$ (theorem 2.4.1), s being the service-time.

The distribution of queueing-time can now be found. If $h(q)$ is its density and $p(n|q)$ is the probability that a departing customer leaves behind him a queue of n customers, conditional on his queueing-time having been q, then

$$p_n = \int_0^\infty p(n|q)\,h(q)\,dq. \tag{12}$$

But $p(n|q)$ is the probability that n customers arrive in a time q, since all n customers must have arrived whilst the customer just departing was queueing: hence it is $P(\lambda q)$ and the corresponding probability generating function is $e^{\lambda q(x-1)}$ (equation 2.6.11). Consequently, if (12) is multiplied by x^n and the equations summed over n, we have (compare the argument leading to (8))

***Theorem* 2.** *In a queueing process of Theorem* 1 *and* $\rho < 1$, $\Phi_q(z)$, *the moment generating function of the queueing-time, is given by*

$$P(x) = \Phi_q[\lambda(x-1)]. \tag{13}$$

In particular the expected queueing-time is

$$\mathscr{E}(q) = \frac{\rho(2-\rho)+\lambda^2\sigma^2}{2\lambda(1-\rho)}. \tag{14}$$

The last result follows on differentiating (13) with respect to x, putting $x = 1$ and using (11).

The distribution of waiting-time, w, can be found by remarking that $q = w+s$, where w and s are independent: hence

$$\Phi_q(z) = \Phi_w(z)\,\Phi(z) \tag{15}$$

by theorem 3.5.4. In particular $\mathscr{E}(q) = \mathscr{E}(w)+\mathscr{E}(s)$, so that

$$\mathscr{E}(w) = \frac{\rho^2+\lambda^2\sigma^2}{2\lambda(1-\rho)}. \tag{16}$$

Discussion of the proof of theorem 1

The theory of the simple queueing process is obviously not adequate for many applications, and the commonest reason for this is that in many situations the service-time distribution is not exponential. The theory of the present section holds for any distribution, but retains the Poisson input and the queue discipline of 'first come, first served'. Important special cases were discussed by Erlang but the general theory here is due to Kendall (1951). The argument used is different from that for the simple queue and it is most important to understand why that argument fails here.

Let us follow through the argument of §4.2, retaining the same notation but replacing the exponential distribution by a general one. We first write down an equation for $p_n(t)$ of the form

$$p_n(t) = \sum_{m=0}^{\infty} p_m(t-\delta t)\, p(n\,|\,m, \delta t), \tag{17}$$

where $p(n\,|\,m, \delta t)$ is the probability of n at time t conditional on m at time $t-\delta t$ (and only the terms $m = n-1$, n and $n+1$ are not $o(\delta t)$). This equation is still true in the general situation of this section. The next stage in the argument is to insert the values of $p(n\,|\,m, \delta t)$. Now as before $p(n\,|\,n-1, \delta t) = \lambda\delta t + o(\delta t)$ because of the Poisson process of arrivals; but we meet a difficulty when we come to $p(n|n+1, \delta t)$. This, apart from terms $o(\delta t)$, is the probability that someone will leave the system, given that the queue is of size $(n+1)$. But this is not given us directly in the specification of the process. When the service-time distribution was exponential we used theorem 4.1.1 to show that it was $\kappa\delta t + o(\delta t)$, irrespective of how long the customer had been with the server. Let us see what happens with a service-time distribution of general density. The necessity part of theorem 4.1.1 shows that it is *only* with the exponential distribution of service-time that the probability of a departure does not depend on how long the customer has been at the service point. With any distribution other than exponential the probability of a departure in an interval of length δt is $\kappa(t_0)\,\delta t + o(\delta t)$, where t_0 is the length of service already completed. So what the specification provides is the probability that a customer will leave the system, given that he has been with the server for a time t_0; whereas what we want for (17) is the probability conditional on the number in the queue. This could be found, by using the generalized addition law, in terms of the probability $\pi_n(t_0|t)$, that, at time t with $(n+1)$ in the queue, the customer being served had been with the server for a time t_0: as

$$p(n|n+1,\delta t) = \int_0^\infty \kappa(t_0)\,\delta t\,\pi_n(t_0\,|\,t)\,dt_0 + o(\delta t), \qquad (18)$$

but $\pi_n(t_0\,|\,t)$ is also unknown. Hence this approach fails. Notice that it fails, not because (17) is false, but because certain of the terms in it are not known.

The argument used in this section avoids the difficulty by considering only the time points when a customer leaves the system. Admittedly not quite the same information is obtained; the probability distribution of queue size is only found at these time points; but this makes little difference in practice. At least it enables the distribution of queueing-time to be obtained (theorem 2). When the queue size is considered at these time points the obvious thing to do is to relate one customer's departure with the departure of the previous customer, using the same type of equation based on the generalized addition law. The result is (4) and the awkward $p(n\,|\,m,\,\delta t)$ of (17) are replaced by the p_{ij} which can be evaluated as shown in terms of known quantities, the π_j. But notice, and again this is important, that this evaluation is only possible because of the Poisson input to the queue. The Poisson process is peculiar in that the probability of an arrival does not depend on any knowledge of earlier arrivals (see the definition in §2.3). With another input process the probability would depend on earlier arrivals and therefore the probability, in (1), that j customers arrive in an interval of time x, might depend on the fact that we knew something about earlier arrivals; namely, that their pattern was such that the customer being served now, arrived to find the server free. Thus (1) would still be valid but the conditional probability occurring in it would be unknown. We could not use theorem 2.3.3 and pass to equation (2). In summary then: the Poisson input and exponential service of the simple queue enable certain conditional probabilities to be written down and the generalized addition law used; when the service-time distribution is general a restriction to departure points of customers enables a modified form of argument to be used but the conditional probabilities need a Poisson input before they can be calculated. The problem with a more general input will not be discussed in this book. It can, however, be reduced to a random walk (§§4.5, 4.6).

Solution of equations for generating functions

The solution of equation (5) using generating functions is a good example of the power of this tool. Notice that $\Pi(x)$ is a *probability* generating function, that is $\sum_j \pi_j = \sum_j p_{0j} = 1$. The difficulty encountered in (7) of one number, p_0, being unknown is common. It can sometimes be met by using the supplementary result $P(1) = 1$, as here, but often a more sophisticated argument is necessary. If $P(x)$ is to be a probability generating function of a non-negative random variable it must be analytic in the region $|x| \leqslant 1$ because it is given there by a convergent power series. Often $P(x)$ is of the form $R(x, p_0)/S(x)$, as it is here, and if $S(x_0) = 0$ with $|x_0| \leqslant 1$ then it must follow that $R(x_0, p_0) = 0$; for otherwise the power series for $P(x)$ would diverge and $P(x)$ would not be analytic. $R(x_0, p_0) = 0$ gives an equation for p_0. If there are several unknowns $p_0, p_1, \ldots$, then $S(x) = 0$ may have several roots $x_0, x_1, \ldots$, each yielding an equation for the unknowns. Typically the number of roots is enough to provide unique solutions for the unknowns. Notice that even when (10) has been obtained it is not obvious that $P(x)$ is a probability generating function. To show that it is we have to show that $P(x)$ is a power series with non-negative coefficients. This can be done and it can be shown that the limits $\lim_{s\to\infty} p_n^{(s)}$ are equal to the coefficients $\{p_n\}$ when $\rho < 1$, but we omit the proof. When $\rho \geqslant 1$ the situation is as in the simple queue and $p_n^{(s)}$ tends to zero for all n and large queues tend to build up.

Ergodicity

The traffic intensity, the ratio of expected service-time to expected interval between arrivals, still plays the critical role in determining the stability of the queueing system and $\rho = 1$ is the critical value. Also p_0 is still equal to $1-\rho$, but now p_0 means the probability that a departing customer begins a slack period and not, as in §4.2, the probability that the server is free. This we saw, by an application of ergodic results, was the proportion of the server's time that was free. However, an ergodic argument shows that this proportion is still $p_0 = 1-\rho$: the argument is as follows. The process from the server's point of view consists of a number of slack periods of lengths $s_1, s_2, \ldots$, mixed with a number of service periods of lengths $s'_1, s'_2, \ldots$. Then by the strong law of large numbers (theorem 3.6.3)

$$\lim_{n'\to\infty}\left(\sum_{i=1}^{n'} s'_i/n'\right) = \kappa^{-1},$$

the mean service-time.† Also, ignoring a small number in the queue, $\sum_{i=1}^{n'} s'_i + \sum_{i=1}^{n} s_i$, where n is the number of slack periods up to the time customer number n' leaves, is the time taken for n' customers to arrive. Hence

$$\lim_{n'\to\infty}\left(\sum_{i=1}^{n'} s'_i + \sum_{i=1}^{n} s_i\right)\Big/ n' = \lambda^{-1},$$

by another application of the strong law, where λ^{-1} is the mean interval between arrivals. Hence

$$\lim_{n'\to\infty}\left(\sum_{i=1}^{n} s_i/n'\right) = \lambda^{-1}-\kappa^{-1}$$

and finally

$$\begin{aligned}\lim_{n'\to\infty}\left\{\left(\sum_{i=1}^{n} s_i\right)\Big/\left(\sum_{i=1}^{n'} s'_i + \sum_{i=1}^{n} s_i\right)\right\} &= (\lambda^{-1}-\kappa^{-1})/\lambda^{-1} \\ &= 1-\rho:\end{aligned}$$

the expression in braces is the proportion of time that the server is idle. Another application of the ergodic principle shows that the unconditional probability of finding the server free is $1-\rho$.

† This, and similar results, require the qualification, 'with probability one'.

Notice that the arguments of this paragraph nowhere use the Poissonian nature of the input, only that the mean interval between arrivals is λ^{-1}. Hence the proportion of time the server is idle is $1-\rho$ whatever be the distribution of these intervals or of the service-times, provided the independence assumptions are retained, though even these might be relaxed.

Whilst using the ergodic principle a further result might be noted. We have just seen that

$$\lim_{n'\to\infty}\left(\sum_{i=1}^{n} s_i/n'\right) = \lambda^{-1}-\kappa^{-1}$$

and by a direct application of the strong law

$$\lim_{n\to\infty}\left(\sum_{i=1}^{n} s_i/n\right) = \lambda^{-1}$$

since (cf. §4.2) the slack periods have an exponential distribution (the Poisson input is needed here). Combining these results

$$\lim_{n\to\infty}(n'/n) = (1-\rho)^{-1}.$$

But n'/n is the ratio of the number, n', of customers served in n busy periods (the numbers of busy periods and slack periods are equal). Hence the average number of customers served in a busy period is $(1-\rho)^{-1}$. The argument of §4.2 still applies to show that the expected length of a busy period is $1/\kappa(1-\rho)$.

Expected values

We have seen that the server's average free time, $1-\rho$, is not affected by the distribution of service-time except through the mean. Equation (11) shows that this is not true of the expected queue size which depends on the variance, σ^2, of the distribution. For a given traffic intensity, the expected queue size is least when $\sigma = 0$, that is when each customer takes the same time, κ^{-1}, to be served. With both λ and κ fixed the same remark holds for the expected queueing-time, equation (14), and the expected waiting-time, equation (16). Hence if the service-time distribution can be controlled it is best to make the time the same for each customer. The expected queue-size is reduced from $\rho/(1-\rho)$ in the exponential case (§4.2) to $\rho(2-\rho)/2(1-\rho)$ and the expected queueing-time from $1/\kappa(1-\rho)$ to $(2-\rho)/2\kappa(1-\rho)$. The case $\sigma = 0$ has been studied by Erlang. Notice that in calculating (11) it was necessary to differentiate *twice*. This is because the value of $P'(1)$ is indeterminate from a single differentiation, just as $P(1)$ was indeterminate from (7), and we had to differentiate to find p_0.

Queueing-time

The argument used to obtain the distribution of queueing-time is different from that used in §4.2. Instead of the arrivals of customers, the departures have been used. Also instead of discussing the joint distribution of n, the number in the queue and q, the queueing-time, through $f(q \mid n)$, the other conditional probability $p(n \mid q)$ has been used (cf. equation 4.2.7 with equation (12)). The earlier argument again fails because theorem 4.1.1 cannot be applied when the service-time distribution is not exponential, the servicing-time remaining of the customer being served will have a distribution dependent on how long his service has already lasted.

Erlang distributions

There is a class of distributions of service-time which is of special interest. Suppose that service consists of a number, n, of, possibly imaginary, consecutive stages, the time of stay in each stage having an $E(\kappa_i)$ distribution, independent of the times of stay in the other stages. The moment generating function of an $E(\kappa_i)$ distribution is, for $z < \kappa_i$,

$$\kappa_i \int_0^\infty e^{zx} e^{-\kappa_i x} dx = \kappa_i/(\kappa_i - z) = (1 - z/\kappa_i)^{-1} \tag{19}$$

so the moment generating function of the service-time distribution is the product of n factors like (19) (theorem 3.5.4), that is

$$\Phi(z) = \prod_{i=1}^{n} (1 - z/\kappa_i)^{-1}. \tag{20}$$

Such a distribution is called an *Erlang* distribution. In particular, if all the κ_i are equal, to κ say, then the distribution is $\Gamma(n, \kappa)$ (theorem 2.3.5). Another way of describing the service-time when the distribution is Erlang is to suppose that service is not complete until n types of incident have occurred; the ith type occurring in a Poisson process of rate κ_i, this Poisson process starting immediately the $(i-1)$st type of incident has occurred. If $\Phi(z)$ is given by (20) then, from (10),

$$P(x) = \frac{(x-1)(1-\rho)}{x\prod_{i=1}^{n}[1-\rho_i(x-1)]-1} = \frac{(1-\rho)}{\prod_{i=1}^{n}(1-\alpha_i x)} \tag{21}$$

for suitable constants, α_i, which can be obtained in terms of the $\rho_i = \lambda/\kappa_i$. Since the probability generating function of $G(\theta)$ (equation 2.1.10) is clearly $(1-\theta)/(1-\theta x)$, the probability generating function of queue size, (21), is that of the sum of n independent geometric distributions, $G(\alpha_i)$. From (13) it follows that the moment generating function of the queueing-time is, with $z = \lambda(x-1)$,

$$\Phi_q(z) = \frac{(1-\rho)}{\prod_{i=1}^{n}\left[1-\alpha_i\left(\frac{z}{\lambda}+1\right)\right]} = \prod_{i=1}^{n}(1-z/\beta_i)^{-1}$$

for suitable β_i, which is again that of an Erlang distribution. With n and the parameters, κ_i, free to be chosen, several distributions can be described to an adequate approximation by an Erlang distribution.

Non-random arrivals

Corresponding to the theory of this section there is a closely similar theory for the case where the service-time distribution is $E(\kappa)$, but the input process is one in which the intervals between successive arrivals are independently distributed according to some common distribution. It is only necessary to consider the queue size when a customer arrives instead of when he leaves. Results similar to theorems 1 and 2 can then be obtained. This may be useful if the customers arrive regularly, as they might if they were given appointments with a fixed interval between appointments and they all arrived on time.

4.4. Renewal theory

Consider a sequence of trials at each of which an event E may, or may not, occur. It will be convenient to suppose that the sequence begins with a trial, number zero, at which E necessarily occurs. A probability measure may be defined on the sequence of trials by giving the probability that E occurs at any trial conditional on the results of the previous trials. Suppose that this probability has the property that, if E_n is the event that E occurs at the nth trial and A_n is any event concerning the occurrence of E at trials *previous* to the nth, then, for $m > 0$,

$$p(E_{n+m} \mid E_n A_n) = p(E_{n+m} \mid E_n) \tag{1}$$

and

$$p(E_{n+m} \mid E_n) = p(E_m \mid E_0) = p(E_m). \tag{2}$$

Equation (1) says that, if E occurs at the nth trial, then the development of the process after the nth trial is independent of what happened before that trial. Equation (2) says that this future development obeys the same probability laws as at the beginning of the process, when E necessarily occurred. Colloquially: whenever E occurs the process starts afresh, like new. Such an event E is called a *recurrent event*. We are interested in how often E occurs in the sequence apart from the zero trial.

Let $p(E_n) = p_n$. The p_n's do *not* form a probability distribution, but, if f_n is the probability that E occurs *for the first time* at the nth trial, then the f_n's form a convergent series since $\sum_{n=1}^{\infty} f_n$ is the probability that E occurs at all in the sequence. Let $\sum_{n=1}^{\infty} f_n = f$. If $f = 1$ the event E is called *persistent* and the sequence $\{f_n\}$ is a probability distribution. If $f < 1$ the event E is called *transient*. A transient event has probability one of occurring only a finite number of times. For the probability of occurring exactly once is $f(1-f)$, since the probability that it will occur is f, the probability that it will not occur again is $1-f$, and these are independent by (1). Similarly, the probability of occurring exactly n times is $f^n(1-f)$, and finally $\sum_{n=0}^{\infty} f^n(1-f) = 1$. A persistent event will certainly occur infinitely often.

It will agree with the language we are using and be mathematically convenient to suppose $p_0 = 1, f_0 = 0$. To avoid tiresome complexities we shall suppose that the greatest common divisor of all n such that $f_n > 0$ is 1. If it is $d(> 1)$ then E can only occur at trials whose numbers are multiples of d and by considering the subsequence of those trials we are reduced to the case $d = 1$. d is called the *period*: if $d = 1$ the event E is called *aperiodic*.

One way in which the sequence can develop so that E occurs at the nth trial, is that E occurs for the first time at the mth trial $(1 \leqslant m \leqslant n)$ and then at the nth trial. The probability of this is

$$\begin{aligned} f_m\, p(E_n \mid E \text{ occurs for first time at } m\text{th trial}) &= f_m\, p(E_n \mid E_m \bar{E}_{m-1} \bar{E}_{m-2} \ldots \bar{E}_1) \\ &= f_m\, p(E_n \mid E_m), \quad \text{by (1)}, \\ &= f_m\, p_{n-m}, \quad \text{by (2)}. \end{aligned}$$

For different m these ways are exclusive and exhaust the possibilities, so, by the addition law and the convention about p_0 and f_0,

$$p_n = \sum_{m=0}^{n} f_m\, p_{n-m} \quad (\text{for } n > 0). \tag{3}$$

The right-hand side is the convolution of the two sequences $\{f_n\}$, $\{p_n\}$ (cf. equation 3.5.5) and this observation suggests the use of generating functions. Let

$$P(z) = \sum_{n=0}^{\infty} p_n z^n, \quad F(z) = \sum_{n=0}^{\infty} f_n z^n. \tag{4}$$

The first certainly converges for $|z| < 1$ since $p_n \leqslant 1$, the second for $|z| \leqslant 1$, and $f = F(1)$. Since (3) holds for $n > 0$, if we multiply it by z^n and sum over n, we obtain on the left-hand side $P(z)$ without the first term $(n = 0)$, and on the right the product $P(z)F(z)$ (cf. theorem 3.5.4). Hence

$$P(z) = [1 - F(z)]^{-1}. \tag{5}$$

Theorem 1. *A necessary and sufficient condition that E be transient is that $\sum_{n=0}^{\infty} p_n$ converges: if $\sum_{n=0}^{\infty} p_n = p$ then*

$$f = (p-1)/p. \tag{6}$$

For any N and $0 < z < 1$

$$\sum_{n=0}^{N} p_n z^n \leqslant P(z) \leqslant \sum_{n=0}^{\infty} p_n,$$

where the last quantity may be infinite. Allowing z to tend to 1, keeping N fixed,

$$\sum_{n=0}^{N} p_n \leqslant \lim_{z\to 1} [1-F(z)]^{-1} \leqslant \sum_{n=0}^{\infty} p_n$$

by (5). If the limit exists it is $(1-f)^{-1}$ and the left-hand inequality shows that the series must converge. If the series converges the two outer expressions are equal and hence the limit exists and has the same value. Hence the theorem is proved with $p = (1-f)^{-1}$, which gives (6).

The sum $\sum_{n=0}^{\infty} nf_n = \mu$, say, is, for a persistent event, the mean of the probability density $\{f_n\}$ and is the expectation of the number of trials up to and including that at which E first occurs; and is therefore (by (2)) the expectation of the number of trials between successive occurrences of E. It may be $+\infty$. If $\mu = \infty$ the recurrent event is called *null*. If $\mu < \infty$ it is called *ergodic*. μ is called the *mean recurrence time* of E.

Theorem 2. *If E is a persistent, aperiodic, recurrent event then*

$$\lim_{n\to\infty} p_n = \mu^{-1}. \tag{7}$$

(*If* $\mu = +\infty$, *the limit is zero.*)

There are two things to prove: first, that the limit exists; secondly, that it is μ^{-1}. The first part, the existence theorem, is given separately at the end of this section. We here assume the limit exists, when it is easy to see what it is.

Let $F_n = \sum_{i=n+1}^{\infty} f_i$, so that $(1-F_n)$ is the distribution function corresponding to $\{f_n\}$ and

$$\mu = \sum_{n=0}^{\infty} F_n, \tag{8}$$

by theorem 2.1.2. Then $f_n = F_{n-1}-F_n$ and (3) may be written

$$p_n = (F_0-F_1)p_{n-1}+(F_1-F_2)p_{n-2}+\ldots+(F_{n-1}-F_n)p_0.$$

Taking all the negative terms on the right over to the left and remembering that E is persistent, so that $F_0 = 1$, we have

$$F_0p_n+F_1p_{n-1}+\ldots+F_np_0$$
$$= F_0p_{n-1}+F_1p_{n-2}+\ldots+F_{n-1}p_0 \quad (n > 0).$$

But if the left-hand side is G_n, say, the right-hand side is G_{n-1}. Hence G_n cannot depend on n, so is equal to the value, $F_0p_0 = 1$, when $n = 1$. Consequently, for all n,

$$G_n = F_0p_n + F_1p_{n-1} + \ldots + F_np_0 = 1. \tag{9}$$

It follows that for any N and $n > N$

$$F_0p_n + F_1p_{n-1} + \ldots + F_Np_{n-N} \leqslant 1.$$

Now, assuming the limit to exist, allow n to tend to infinity keeping N fixed. Then, only a finite number of terms being involved,

$$\lim_{n\to\infty} p_n \leqslant \left(\sum_{i=0}^{N} F_i\right)^{-1} \tag{10}$$

for all N. By (8), if $\mu = \infty$, this proves the result.

We need only consider the case $\mu < \infty$. It is then possible, for any $\epsilon > 0$, to find N such that $\sum_{i=N+1}^{\infty} F_i < \epsilon$, and, with the fact that $p_n \leqslant 1$, obtain the result from (9) that

$$F_0p_n + F_1p_{n-1} + \ldots + F_Np_{n-N} + \epsilon \geqslant 1.$$

Again assume the limit to exist, allow n to tend to infinity keeping N fixed, and we obtain

$$\lim_{n\to\infty} p_n \geqslant (1-\epsilon)\left(\sum_{i=0}^{N} F_i\right)^{-1}. \tag{11}$$

(8), (10) and (11) establish the result since ϵ is arbitrary.

Recurrent events

In previous sections the case of *independent* trials with *constant* probability of success has been considered. The results of the present section apply to trials without either of these restrictions. The assumption of independence is replaced by a weaker one in which it is merely assumed that whenever E occurs, the past and future trials are independent (equation (1)). The assumption of constant probability of success is weakened to make the probabilities revert to their original values whenever E occurs (equation (2)). Both characteristics of the earlier case are therefore preserved, but in much weakened forms. In particular, nothing is said about the development of the trials when E does not occur. Furthermore, the assumptions might be appropriate when considering trials, each of which results in the values of one or several random variables. Provided only an event E can be defined in terms of the random variables satisfying (1) and (2) then the theorems may be applied.

Example 1. *Industrial renewal theory.* Consider a piece of industrial equipment liable to failure. The trials are the separate occasions on which it is used, and if, on one of these, it fails, it is replaced by a new piece. Let E be the event of failure. Since a new piece is used whenever E occurs it will be reasonable to assume its performance independent of the other pieces that have been used, and (1) obtains. If the pieces are taken randomly from a stock they will have similar probability characteristics, and (2) obtains. (The process is assumed to begin with a new piece of equipment.)

Here p_n is the probability that a failure will occur on the nth occasion that the equipment is used: f_n is the probability that a new piece will fail on the nth occasion that it is used. $1-f$ is the probability that the piece will last for ever. If $f = 1$ replacement is eventually certain and μ is the average lifetime of the piece. For large n, theorem 2 says the chance of a failure on the nth occasion that a piece of equipment is used is μ^{-1}.

In this particular example it is often more convenient to define the probability structure in terms of λ_n, the probability that a piece, having been used without failure on $(n-1)$ occasions will fail on the nth: that is, the probability that a piece of known 'age' will fail. The reason being that the age will normally be known and therefore this probability is the practically relevant one. Clearly, if $f = 1$,

$$\lambda_n = p(E_n \mid \bar{E}_{n-1}\, \bar{E}_{n-2} \ldots \bar{E}_1)$$

$$= p(E_n \bar{E}_{n-1} \ldots \bar{E}_1)/p(\bar{E}_{n-1} \ldots \bar{E}_1),$$

hence $$\lambda_n = f_n \bigg/ \left(1 - \sum_{i=1}^{n-1} f_i\right) = f_n/F_{n-1}. \tag{12}$$

Conversely $$f_n = \lambda_n \prod_{i=1}^{n-1} (1-\lambda_i). \tag{13}$$

An important special case is where λ_n is constant, equal to λ, say; so that the probability of failure does not depend on the age of the piece. From (13) $f_n = \lambda(1-\lambda)^{n-1}$, so that ultimate failure is certain and the distribution of $(n-1)$ is geometric. We easily obtain

$$F(z) = \lambda z[1-(1-\lambda)z]^{-1} \quad \text{and} \quad P(z) = 1+\lambda z/(1-z)$$

so that $p_n = \lambda$ for all $n > 0$, as is directly obvious. Conversely, if the distribution of $(n-1)$ is geometric then λ_n is constant. For many pieces of equipment λ_n will, of course, increase with n.

Example 2. *Simple random walk.* Consider the simple random walk (§2.5). This is a sequence of trials with each of which is associated the position of the particle, x_n, so that the outcomes of the trials are not independent. Let E be the event that the particle returns to the origin. Then E_n is the event, $x_n = 0$. We show that E is a recurrent event, but first we notice that E can only occur at trials with even numbers: that is, it has period 2. We shall therefore suppose that n is always even and only consider every other trial. If $n = 2m$, m will play the role in the theory previously occupied by n. If $x_n = 0$ the future depends on $u_{n+1}, u_{n+2}, \ldots$ (equation 2.5.1), whereas the past depended on $u_1, u_2, \ldots, u_n$: the u's are independent, so (1) is satisfied. All the u's have the same distribution so (2) is satisfied. Notice that the event, $x_n = 0$, is not peculiar: for any integer c the event of return to c is similarly a recurrent event.

From equation 2.5.2, with $s = 0$, $n = 2m$,

$$p_m = \binom{2m}{m}(pq)^m, \tag{14}$$

and to decide whether E is persistent or not we have to consider the convergence of Σp_m, for which we need the behaviour of (14) for large m. By Stirling's formula,

$$m! \sim \sqrt{(2\pi)}\, e^{-m} m^{m+\frac{1}{2}}, \tag{15}$$

we easily obtain that

$$p_m \sim \frac{1}{\sqrt{(\pi m)}}(4pq)^m. \tag{16}$$

If $p \neq \frac{1}{2}$, $4pq < 1$ and the series converges faster than the geometric series with common ratio less than 1. Hence Σp_m

converges and, by theorem 1, E is transient. If $p = \frac{1}{2}$, $4pq = 1$ and the series behaves like $\Sigma m^{-\frac{1}{2}}$ which diverges and E is persistent. However, then $p_m \to 0$, so that the mean recurrence time is infinite. Hence if the walk has a drift ($p \neq q$) it will return to the origin at most a finite number of times: if it has no drift ($p = q$) then return is certain, and it will return infinitely often, but the expected time until the return is infinite. In the case of drift ($p \neq q$) f, the probability that it will at any time return to the origin can be found using theorem 1:

$$\sum_{m=1}^{\infty} p_m = (1-4pq)^{-\frac{1}{2}},$$

from (14). But since $p+q = 1$, $(1-4pq)^{-\frac{1}{2}} = |p-q|^{-1}$. Hence, from (6),

$$f = 1-|p-q|.$$

The case $p = \frac{1}{2}$ solves the problem mentioned in the introduction to chapter 1 of how long one can expect to go on tossing a fair penny until the numbers of heads and tails are equal.

This example can easily be generalized to higher numbers of dimensions. Thus, in two dimensions, if the walk can proceed from any point (x, y) with integer co-ordinates to $(x \pm 1, y \pm 1)$ with equal probabilities of $\frac{1}{4}$, so that there is no drift, it is equivalent to two independent simple random walks on the x- and y-axes, both without drift. It can only return to the origin if both the x- and y-co-ordinates are zero, and hence the probability of return to the origin is p_m^2, where p_m is given by (14). Σp_m^2 behaves like Σm^{-1}, which still diverges, so that the walk is certain to return to the origin, but the recurrence time is infinite. In s dimensions we deal with Σp_m^s which behaves like $\Sigma m^{-\frac{1}{2}s}$ and converges for $s > 2$. Hence, in dimensions higher than two, return to the origin is not certain and that event is transient.

Other events in connexion with the random walk may be similarly studied. For example, one is sometimes interested in 'records'; that is, positions of the walk, x_n, which are such that $x_n > x_i$ for all $i < n$. If E is the event of a record then it is easy to see that it is a recurrent event. Equation (1) follows because, if x_n is a record, whether or not E occurs at trial $(n+m)$ depends only on $x_{n+m}-x_n$. Equation (2) follows since the distribution of $x_{n+m}-x_n$ is the same as that of $x_m - x_0 = x_m$. Another example concerns 'success' runs. If E is the event of r, say, successive steps of $+1$ (corresponding to a 'success' at a particular trial and the $(r-1)$ previous ones), it is easy to see that this is a recurrent event, and one can calculate the frequency with which such runs occur.

Extension to continuous time

The main applications of renewal theory are to Markov processes to be studied in the next section. Meanwhile we make a few general comments. The idea of a recurrent event has been formulated for sequences, but it extends without difficulty to general stochastic processes in continuous time. We shall not describe the generalization, but note that the concept has been used in the treatment of the queue with Poisson input in §4.3. The development of that process from any instant in time depends on the behaviour before that instant because the probability of someone leaving the system depends on how long he has been with the server. At those instants of time when a customer leaves the system, however, the future development does not, because of the Poisson input, depend on the past. Therefore if the queue starts at $t = 0$ with j customers in the queue and someone just having left, the event $E^{(j)}$ of someone leaving with j customers behind him is a recurrent event, in the sense that the process develops independently of the past according to the same laws as held at $t = 0$. These instants were just those used in §4.3 and the analysis there is an example of the fruitful use of recurrent events to study a process.

Ergodicity

Theorem 2 is another example of an ergodic theorem (§3.6). Consider a long sequence of trials. The average interval between the occurrences of E will be μ so that on the average E will occur on one out of every μ trials. Consequently, over a long period of time the proportion of times E occurs is μ^{-1}. This, by the theorem, is equal to the probability that, at a particular trial, a long way distant from the influence of the start, E will occur:

that is, the long-run frequency of E at a particular trial in repeated sequences. Hence an average over one sequence equals the average at a fixed trial over several sequences.

Distribution theory

In virtue of the central limit theorem (theorem 3.6.1) we can say more about the distribution of the number of occurrences of E. Let N_t be the number of occurrences of E in the first t trials. Let x_i be the number of trials after the $(i-1)$st occurrence up to and including the ith—the lifetime of the ith piece in industrial renewal terminology. Then, for fixed n and t, $N_t \geqslant n$ iff $x_1+x_2+\ldots+x_n \leqslant t$. Hence

$$p(N_t \geqslant n) = p\left(\sum_{i=1}^{n} x_i \leqslant t\right). \tag{17}$$

But the x_i are independent with a common distribution. This distribution has mean μ. Suppose $\mu < \infty$ and that also the variance, σ^2, is finite. Then by the central limit theorem, for any fixed z,

$$\lim_{n\to\infty} p\left[\left(\sum_{i=1}^{n} x_i - n\mu\right)\Big/\sigma\sqrt{n} < z\right] = \Phi(z), \tag{18}$$

the distribution function of the normal law. Consequently from (17) and (18)

$$\lim_{n\to\infty} p(N_t \geqslant n) = \Phi(z), \tag{19}$$

where

$$(t-n\mu)/\sigma\sqrt{n} = z.$$

This last equation gives $\sqrt{n} = \{-z\sigma+(z^2\sigma^2+4\mu t)^{\frac{1}{2}}\}/2\mu$, on solving for $\sqrt{n}$ which must be positive. Since n and $t\to\infty$, z fixed, this gives

$$n = t\mu^{-1} - z\sigma t^{\frac{1}{2}}\mu^{-\frac{3}{2}} + O(1).$$

Substitution of this into (19), to eliminate n, gives

$$\lim_{t\to\infty} p\left(\frac{N_t - t\mu^{-1}}{\sigma t^{\frac{1}{2}}\mu^{-\frac{3}{2}}} \leqslant z\right) = \Phi(z). \tag{20}$$

Equation (20) says that N_t is asymptotically normally distributed with mean $t\mu^{-1}$ and variance $\sigma^2 t/\mu^3$. Notice that this does not prove that for large t the mean and variance of N_t are approximately $t\mu^{-1}$ and $\sigma^2 t/\mu^3$ respectively, though this is true.

That the approximate mean is $t\mu^{-1}$ follows from theorem 2. For, by considering a variable which is 1 when E occurs and 0 otherwise, it is clear that $\mathscr{E}(N_t) = \sum_{n=1}^{t} p_n$. Hence $\mathscr{E}(N_t/t)$ behaves like $t^{-1} \sum_{n=1}^{t} p_n$ which has limit μ^{-1}. The proof that the variance is as stated will not be given.

If (20) is applied to the industrial renewal process above with $\lambda_n = \lambda$ we have $\mu = \lambda^{-1}$, $\sigma^2 = (1-\lambda)\lambda^{-2}$ and hence

$$\lim_{t\to\infty} p\left(\frac{N_t - \lambda t}{[\lambda(1-\lambda)t]^{\frac{1}{2}}} \leqslant x\right) = \Phi(x),$$

which agrees with the fact that, the trials being now independent with $p_n = \lambda$, N_t is binomial with mean λt and variance $\lambda(1-\lambda)t$, and therefore asymptotically normal (corollary to theorem 2.5.2).

Delayed recurrent events

A trivial, but useful, extension of renewal theory is possible. It is not necessary to suppose that E occurs at a fictitious zero trial. We can replace (2) by the weaker condition that $p(E_{n+m} \mid E_n)$ does not depend on n, still calling it p_m, and let the process start in any way we like. E is then called a *delayed* recurrent event. Let this starting condition be defined in terms of f_n^*, the probability that E occurs for the first time at the nth trial: after E has occurred the process continues to develop according to the p's. If p_n^* is the probability that E occurs at the nth trial in this new process, then clearly, as before

$$p_n^* = f_n^* + f_{n-1}^* p_1 + \dots + f_1^* p_{n-1}.$$

Since Σf_n^* necessarily converges, to f^* say, and p_n tends to a limit,† μ^{-1}, it follows as in the argument from equation (9) onwards in the proof of theorem 2 that $\lim_{n\to\infty} p_n^* = f^*\mu^{-1}$. In particular if E is certain to occur at least once, $f^* = 1$, and $\lim_{n\to\infty} p_n^* = \mu^{-1}$. Therefore it does not matter for the ergodic property of the process how it starts provided only that the event E is certain to occur at least once.

† If E is transient then $p_n \to 0$ since Σp_n converges (theorem 1).

Existence theorem

We now prove the existence of the limit stated in theorem 2. Since $\{p_n\}$ is a bounded sequence it has an upper limit, $\overline{\lim\limits_{n\to\infty}}\, p_n = \pi$, say, and a subsequence $\{p_{n_k}\}$ such that $\lim\limits_{k\to\infty} p_{n_k} = \pi$. Let s be any integer such that $f_s > 0$. Then from equation (3)

$$\pi = \lim_{k\to\infty} p_{n_k} = \lim_{k\to\infty} \left\{ f_s p_{n_k-s} + \sum_{\substack{i=1\\ i\neq s}}^{n_k} f_i p_{n_k-i} \right\}.$$

Because of the convergence of $\sum\limits_{i\neq s} f_i (= 1-f_s)$ the sum in the braces may be treated as a sum up to a fixed N plus an arbitrarily small quantity (compare the way in which (11) was obtained) and hence

$$\pi \leqslant f_s \underline{\lim_{k\to\infty}}\, p_{n_k-s} + (1-f_s)\overline{\lim_{n\to\infty}}\, p_n = f_s \underline{\lim_{k\to\infty}}\, p_{n_k-s} + (1-f_s)\pi$$

and consequently, since $f_s \neq 0$, $\underline{\lim\limits_{k\to\infty}}\, p_{n_k-s} \geqslant \pi$. This implies, by the definition of π, that $\lim\limits_{k\to\infty} p_{n_k-s} = \pi$.

Hence we have proved that if $\lim\limits_{k\to\infty} p_{n_k} = \pi$, then $\lim\limits_{k\to\infty} p_{n_k-s} = \pi$ for any s such that $f_s > 0$. If $f_1 > 0$ we may apply this result with $s = 1$ repeatedly to show that $\lim p_{n_k-s} = \pi$ for all s and hence $\lim\limits_{n\to\infty} p_n = \pi$, as we had to prove. If $f_1 = 0$ then we have to use a result in elementary number theory that says that any sufficiently large number can be written in the form $\sum\limits_{i=1}^{m} c_i s_i$, where $s_1, \ldots, s_m$ are a finite set of numbers such that $f_{s_i} > 0$, because the greatest common divisor of all s with $f_s > 0$ is 1 (aperiodic case). Then $\lim\limits_{k\to\infty} p_{n_k - \Sigma c_i s_i} = \pi$ by repeated applications of the result and hence $\lim\limits_{n\to\infty} p_n = \pi$ as required, since $\sum\limits_{i=1}^{m} c_i s_i$ is arbitrary.

4.5. Markov chains

Consider an enumerable number of elements, E_i ($i = 0, 1, \ldots$) called *states*; and a sample space, each elementary event of which is an infinite sequence $(x_0, x_1, x_2, \ldots)$ of states: that is, each x_j is some E_i. If x_i is thought of as coming before x_{i+1} in time, then we have a stochastic process which moves from state to state: the movement from one state to the next is called a *transition* or a *step*. A probability distribution over this sample space can be defined in terms of the density of x_n conditional on the states assumed by $(x_0, x_1, \ldots, x_{n-1})$. The process is said to be *Markov* if, for $n > 0$, this density depends only on the state assumed by x_{n-1}. In symbols: the process is Markov if, whenever A is any event depending only on $x_0, x_1, \ldots, x_{n-2}$

$$p(x_n = E_j \mid x_{n-1} = E_i, A) = p(x_n = E_j \mid x_{n-1} = E_i) \qquad (1)$$

for all i, j. It follows that a Markov process is defined by (1) and $p(x_0 = E_i)$. The probability in (1) is the *transition probability* of the transition from E_i to E_j. A Markov process is *temporally homogeneous* if (1) does not depend on n. In this case the transition probability (1) is written p_{ij} and we write $p(x_n = E_i) = p_i^{(n)}$. $\mathbf{p}_n$ denotes the column vector of elements $p_i^{(n)}$ ($i = 0, 1, \ldots$) and $\mathbf{P}$ is the matrix of elements p_{ij}, called the *transition matrix*.† We shall refer to a temporally homogeneous Markov process with an enumerable number of states as a *Markov chain*. If the number of states is finite, the chain will be called *finite*.

The basic equation for studying Markov chains is easily obtained from the generalized addition law (theorem 1.4.4):

$$\begin{aligned} p_j^{(n)} &= p(x_n = E_j) \\ &= \sum_i p(x_{n-1} = E_i)\, p(x_n = E_j \mid x_{n-1} = E_i) \\ &= \sum_i p_i^{(n-1)} p_{ij}, \end{aligned}$$

or, in matrix notation,

$$\mathbf{p}_n' = \mathbf{p}_{n-1}'\mathbf{P}. \qquad (2)$$

† The order of the suffixes is important. Here p_{ij} refers to a transition from E_i to E_j. Some authors put the suffixes in the reverse order and p_{ij} refers to a transition from E_j to E_i.

Repeated use of this result establishes that

$$\mathbf{p}_n' = \mathbf{p}_0'\mathbf{P}^n, \tag{3}$$

so that the probability distribution, $\mathbf{p}_n$, is obtained in terms of the known $\mathbf{p}_0$, the initial probability distribution, and the known transition matrix. The elements of $\mathbf{P}^n$ will be denoted by $p_{ij}^{(n)}$. A further use of the generalized addition law shows that

$$\begin{aligned} p_{ij}^{(2)} &= p(x_n = E_j | x_{n-2} = E_i) \\ &= \sum_k p(x_n = E_j \,|\, x_{n-1} = E_k) p(x_{n-1} = E_k | x_{n-2} = E_i), \end{aligned}$$

and in general that

$$p_{ij}^{(n)} = p(x_n = E_j \,|\, x_0 = E_i).$$

Similarly, $$p_{ij}^{(n)} = p(x_{m+n} = E_j \,|\, x_m = E_i), \tag{4}$$

the probability of a transition from E_i to E_j in n steps irrespective of m. Of course, $p_{ij}^{(1)} = p_{ij}$.

Suppose the Markov chain starts in a particular state, say E_i: that is, $x_0 = E_i$. Then, by considering merely whether or not the chain is, after any transition, in state E_i it is clear by a comparison of (1) with equation 4.4.1 and of (4) with equation 4.4.2 that the event of being in state E_i is a recurrent event. Indeed, a Markov chain can be defined as a stochastic process on a number of events, each of which is recurrent in this sense. Extensive use of the results of §4.4 can therefore be made in the study of Markov chains, and descriptions of the recurrent event of being in state E_i can be transferred to descriptions of the state E_i itself. In particular, every state can be described as either *persistent* or *transient* according as the recurrent event of being in that state is persistent or transient. Similarly, the state will have a *period*, and if persistent will have a *mean recurrence time*: if μ_i say, is infinite the state will be called *null*, otherwise it is *ergodic*. In theorem 4.4.1 we saw that the character of the event is determined by the behaviour of a series: the relevant series here is $\sum_n p_{ii}^{(n)}$; E_i is persistent or transient according as this diverges or converges.

Suppose that E_i is persistent but the chain starts in another state E_j. Then it is clear that the event of being in E_i will be a delayed recurrent event and, from the remarks in §4.4, the ergodic properties of E_i will remain the same provided that the event E_i is certain to occur at least once. It is therefore natural to study whether a chain starting in E_j will reach E_i and vice versa. If there exists some m such that $p_{ij}^{(m)} > 0$ then we say that E_j can be *reached* from E_i because it is possible to pass from E_i to E_j in m steps. If E_j can be reached from E_i and E_i from E_j in some finite number of steps, we say that E_i and E_j are *connected*.

Suppose E_i is persistent and E_j can be reached from E_i. Since E_i is persistent, if the chain reaches E_j it must be certain to return from E_j to E_i so that E_i and E_j are connected. Hence there exists n such that $p_{ji}^{(n)} > 0$. Now if $c = p_{ij}^{(m)} p_{ji}^{(n)}$, then $c > 0$, and

$$p_{ii}^{(m+n+s)} \geqslant p_{ij}^{(m)} p_{jj}^{(s)} p_{ji}^{(n)} = c p_{jj}^{(s)} \tag{5}$$

since one way of returning to E_i in $(m+n+s)$ steps is to pass to E_j in m, return to E_j s steps later, and return to E_i in n further steps. Similarly,

$$p_{jj}^{(m+n+s)} \geqslant p_{ji}^{(n)} p_{ii}^{(s)} p_{ij}^{(m)} = c p_{ii}^{(s)}. \tag{6}$$

Equations (5) and (6) show that the series $\sum_n p_{ii}^{(n)}$ and $\sum_n p_{jj}^{(n)}$ behave alike and therefore E_i being persistent means that E_j is also persistent. If E_i is null, $p_{ii}^{(s)} \to 0$ and, by (5), $p_{jj}^{(s)} \to 0$, so that E_j is also null, and conversely, by (6). Furthermore, if E_i has period d, then with $s = 0$ in (5) $m+n$ must be a multiple of d, and hence if $p_{jj}^{(s)}$ exceeds zero for s not a multiple of d, again by (5), $p_{ii}^{(m+n+s)}$ would exceed zero for $m+n+s$ not a multiple of d, which is impossible; therefore E_j also has period d.

These arguments prove that the states that can be reached from any persistent state are all persistent with the same period. They thus form a *closed* set of states: once the chain is in the closed set it cannot leave it. Furthermore, E_i being persistent, the chain will return to E_i infinitely often: each time it does so there is a probability $p_{ij}^{(m)} > 0$ that it will pass to E_j in m steps. The probability that it will never pass from E_i to E_j in m steps is therefore $\lim_{n\to\infty} (1-p_{ij}^{(m)})^n = 0$. Hence not only *can* E_j be reached from E_i; it is *certain* to be reached; and similarly E_i from E_j. We therefore have

Theorem 1. *In any Markov chain the states may be divided into closed sets of persistent states, all the states of any one closed set having the same period and being all null or all ergodic; and transient states. From the states of any closed set only states of that set can and will be reached.*

As in §4.4, we henceforth consider only the aperiodic case, $d = 1$. If $d > 1$ in a closed set it suffices to consider the process $(x_0, x_d, x_{2d}, \ldots)$ instead of the original one.

Before giving the next proof we remark that if Σz_i is an absolutely convergent series (in particular if $\{z_i\}$ is a probability density) and $|u_i^{(n)}| \leqslant 1$, then

$$\lim_{n\to\infty} \Sigma z_i u_i^{(n)} = \Sigma z_i \lim_{n\to\infty} u_i^{(n)}, \tag{7}$$

provided either side exists.

Theorem 2. *In a Markov chain containing only one closed aperiodic set of persistent states, and no transient states, either*

(*a*) *every state is ergodic and for all* $\mathbf{p}_0$

$$\lim_{n\to\infty} p_i^{(n)} = p_i = \mu_i^{-1}. \tag{8}$$

Furthermore $$\sum_i p_i = 1 \quad \text{and} \quad \mathbf{p}' = \mathbf{p}'\mathbf{P}, \tag{9}$$

where $\mathbf{p}$ *is the column vector of elements* p_i *forming the* stationary distribution *of the chain.* $\mathbf{p}$ *is the unique solution of* (9) *with* $\Sigma p_i = 1$. *Or*

(*b*) *every state is null and for all* $\mathbf{p}_0$

$$\lim_{n\to\infty} p_i^{(n)} = 0 \tag{10}$$

and there is no stationary distribution.

Every state is either null or ergodic by theorem 1. From the basic renewal theorem 4.4.2 it follows that $\lim_{n\to\infty} p_{ii}^{(n)} = \mu_i^{-1}$. Hence if the chain starts in E_i, $\lim_{n\to\infty} p_i^{(n)} = \mu_i^{-1} = p_i$, say. But if it starts in E_j it is certain to reach E_i by theorem 1, because there is only one closed set, and therefore the ergodic property persists and $\lim_{n\to\infty} p_{ji}^{(n)} = p_i$. For general $p_i^{(0)}$,

$$p_i^{(n)} = \sum_j p_j^{(0)} p_{ji}^{(n)}$$

(equation (3)) and, by the remark (7),

$$\lim_{n\to\infty} p_i^{(n)} = \sum_j p_j^{(0)} p_i = p_i.$$

This proves (8) and (10).

Now

$$p_i^{(n+1)} = \sum_j p_j^{(n)} p_{ji} \geqslant \sum_J p_j^{(n)} p_{ji}$$

for any finite set J of integers j. Hence allowing n to tend to infinity $p_i \geqslant \sum_J p_j p_{ji}$ and therefore $p_i \geqslant \sum_j p_j p_{ji}$. But $\sum_J p_{ij}^{(n)} \leqslant 1$, so $\sum_J p_j \leqslant 1$ and hence $\sum_j p_j \leqslant 1$: so that on summing the result of the last sentence over i and reversing the order of summation of the absolutely convergent double series, we have $\sum_i p_i \geqslant \sum_j p_j$, since $\sum_i p_{ji} = 1$. Consequently the inequality becomes an equality and $p_i = \sum_j p_j p_{ji}$ which is (9). In the null case such a stationary distribution (satisfying (9) with $\Sigma p_i = 1$) cannot exist since if $p_i^{(0)} = p_i$ then $p_i^{(n)} = p_i$ for all i contradicting (10). It remains only to prove that in the ergodic case ($p_i > 0$) the p_i form a distribution and are the unique solutions of (9). We know Σp_i converges to a *positive* (non-zero) sum so that there exists one solution, namely $p_i/\Sigma p_i$, of $z_i = \sum_j z_j p_{ji}$ with $\Sigma z_i = 1$. Also any solution of $z_i = \sum_j z_j p_{ji}$ satisfies $z_i = \sum_j z_j p_{ji}^{(n)}$ and allowing n to tend to infinity, using (7), $z_i = \sum_j z_j p_i = p_i$. Hence there is a unique solution, namely $z_i = p_i$, $\Sigma z_i = \Sigma p_i = 1$. The distribution is stationary since if $\mathbf{p}_0 = \mathbf{p}$ then $\mathbf{p}_n = \mathbf{p}$ for all n. The process already being temporally homogeneous it will, with $\mathbf{p}_0 = \mathbf{p}$, be stationary (§4.1). This completes the proof.

We have proved that, if all states are null, a stationary distribution does not exist. The same proof establishes the same

result if all states are transient. Consequently we have the
Corollary. *If all the states are connected, there exists a stationary distribution iff they are all ergodic.*

Theorem 3. *A finite Markov chain cannot forever stay in transient states, and if there is just one closed set then there exists a unique stationary distribution.*

If E_i is transient then, in any realization apart from those in a set having zero probability, there exists an n_i such that the process is never in E_i after the n_ith transition. Since the number of states is finite, the process is never in a transient state after $n = \max n_i$ transitions. If there is one closed set it is certain to enter it. Since $\sum_i p_{ji}^{(n)} = 1$, the summation being over the *finite* number of states in the closed set, $\Sigma p_i = 1$ and hence not all states can be null. They must therefore be all ergodic and theorem 2(*a*) applies.

Examples

We give some examples of Markov chains. We have already met one in the simple random walk (§2.5) and we first consider this in various forms.

(*a*) *No barriers.* This is the form considered in §2.5. The states are the integer values on the real line and the Markov property (equation (1)) clearly obtains since the transition depends only on the u_i. We saw in §4.4 that the event of being at the origin was a recurrent event which was persistent only when $p = q$ and was otherwise transient. The origin has no special position in this long-run property and therefore this result applies to the event of being in any state. Hence if $p = q$ all states are persistent and each is obviously connected to every other: there is therefore one closed set. If $p \neq q$ all states are transient. When $p = q$ we saw (§4.4) that the recurrence time was infinite and therefore all states are null. Consequently there is no stationary distribution. This agrees with the result of §2.5 where we found that the distribution after a finite number of transitions was binomial with variance which increased with this number. Notice that the simple walk has period 2.

(*b*) *Two absorbing barriers.* As explained in §2.5 the usual applications of simple random walk theory (games of chance, sampling inspection, etc.) have barriers present. We first consider the case where there are two absorbing barriers—absorbing in the sense that when the walk reaches either of them it remains there forever: the walk stops. Specifically let the walk take place on the integers 0, 1, 2, ..., N, and let E_i denote the state of being at integer point i. Then the transition matrix, if the states E_0 and E_N are absorbing, will have the form

$$\begin{pmatrix} 1 & 0 & 0 & \dots & \dots & \dots & 0 \\ q & 0 & p & 0 & \dots & \dots & 0 \\ 0 & q & 0 & p & 0 & \dots & 0 \\ 0 & 0 & q & 0 & p & 0 & 0 \\ \dots & \dots & \dots & \dots & \dots & \dots & \dots \\ 0 & \dots & \dots & 0 & q & 0 & p \\ 0 & \dots & \dots & \dots & 0 & 0 & 1 \end{pmatrix}, \tag{11}$$

where p_{ij} $(i, j = 0, 1, ..., N)$ is the probability of transition from E_i to E_j. Clearly E_0 and E_N form two closed sets of persistent states. (A closed set consisting of a single state is often called *absorbing*.) $E_i (i \neq 0, N)$ cannot be persistent for if it were, since E_0 can be reached from E_i, by theorem 1, E_i could be reached from E_0, which is impossible. They must therefore be transient, and by theorem 3 the walk must eventually finish up in either E_0 or E_N (the walk must be absorbed in one of the barriers). In the next section we shall see what the probability is that it will be absorbed in, say, E_0.

(*c*) *Two impenetrable barriers.* In some situations the barriers (still at E_0 and E_N) are *impenetrable*, in the sense that a step beyond them is thwarted and a particle stays at the barrier instead, but if the step is towards the other barrier it is taken in the normal way. The transition matrix is now

$$\begin{pmatrix} q & p & 0 & \dots & \dots & \dots & \dots & 0 \\ q & 0 & p & 0 & \dots & \dots & \dots & 0 \\ 0 & q & 0 & p & 0 & \dots & \dots & 0 \\ 0 & 0 & q & 0 & p & 0 & \dots & 0 \\ \dots & \dots & \dots & \dots & \dots & \dots & \dots & \dots \\ 0 & \dots & \dots & \dots & 0 & q & 0 & p \\ 0 & \dots & \dots & \dots & \dots & 0 & q & p \end{pmatrix}. \tag{12}$$

In this situation it is clear that all states are connected and therefore, by theorem 3, they cannot all be transient so they must be ergodic and there exists a stationary distribution. This is easily found since it satisfies (9) with **P** given by (12). The equations are (cf. equation 2.5.3)

$$\left.\begin{aligned} p_0 &= p_0 q + p_1 q \\ p_i &= p_{i-1} p + p_{i+1} q \\ \text{and} \qquad p_N &= p_{N-1} p + p_N p \end{aligned}\right\} \quad (0 < i < N). \tag{13}$$

These are easily solved when $p \neq q$ either by remarking that the equations for $0 < i < N$ form a set of second-order difference equations with subsidiary equation $x = p + x^2 q$ with roots (p/q) and 1, so that $p_i = A(p/q)^i + B$, where A and B can be found from the other two equations and $\Sigma p_i = 1$: or by solving iteratively giving $p_1 = p_0(p/q)$, $p_2 = p_1(p/q)$, and generally $p_i = p_0(p/q)^i$. Hence, either way,

$$p_i = \left(\frac{p}{q}\right)^i \left\{\frac{1-p/q}{1-(p/q)^{N+1}}\right\}. \tag{14}$$

The stationary distribution is therefore a truncated geometric one with mode at E_0 if $p < q$ and at E_N if $p > q$. If $p = q$ all states are equally likely: this can be seen either by a limiting process, $p \to \frac{1}{2}$, applied to (14) or directly by solving (13).

(*d*) *One absorbing barrier*. Let E_0 be absorbing but otherwise the walk is unrestricted; there are an infinity of states $E_0, E_1, \ldots$. The infinite transition matrix is as (11) without the bounds at the right or the bottom. An important way in which an infinite chain can differ from a finite one is that it can stay forever in transient states. (By theorem 3 this is not possible for a finite one.) Here E_0 is persistent but all other states are transient by the same argument as was used in (*b*). We prove in the next section that if $p \leqslant q$ the walk is certain to reach E_0 and there exists a unique stationary distribution, namely $p_0 = 1$, $p_i = 0$, $i > 1$, but that if $p > q$ there is a non-zero probability that the walk will forever remain in transient states and so never be absorbed.

(*e*) *One impenetrable barrier*. Let E_0 be an impenetrable barrier but otherwise the walk is unrestricted; there are an infinity of states $E_0, E_1, \ldots$. The infinite transition matrix is as (12) without the bounds at the right or the bottom. Here all states are connected so that if one is persistent all are. It is possible to distinguish the ergodic case from other possibilities by using the corollary to theorem 2 and seeing when a stationary distribution exists. Such a distribution satisfies the first equation of (13) and the second equation for all $i > 0$. It is easy to see as before that the only solution is $p_i = A(p/q)^i + B$ and hence a stationary distribution exists iff $p < q$. Hence when $p < q$ the chain is ergodic and the stationary distribution is geometric with $p_i = (q-p)(p/q)^i/q$. When $p \geqslant q$ either all states are null, or all transient. We show how these can be distinguished in the next section.

(*f*) *Queueing theory*. We have already met a Markov chain in the queueing process of §4.3. Instead of considering the process in its natural form in continuous time we discussed only the time points when a customer leaves the system. The states E_j are the states of leaving j customers behind at the end of service ($j = 0, 1, \ldots$). Equation 4.3.3 gives the transition matrix and equation 4.3.4 is equivalent to (2) above. In that section we immediately passed in equation 4.3.5 to the equation for the stationary distribution, (9) above. It is clear that all states are connected. A stationary distribution only exists if $\rho < 1$ and hence, as in (*e*), the chain is ergodic (and the limit exists—a fact which was not proved in §4.3) if $\rho < 1$. If $\rho \geqslant 1$ all the states are either null or transient. As in example (*e*) we shall see how to distinguish these two cases in the next section.

These examples cover the main types that can occur. It is also instructive to have an example of a non-Markov chain. Consider a retailer selling articles like refrigerators. Let E_i be the state of having i refrigerators in stock ($i = 0, 1, \ldots, N$), N being the capacity of his stockroom, and consider the stock on successive days. The stock is reduced by sales and it might be reasonable to suppose that the day-by-day sales are independent so that a transition from E_i to E_j ($j \leqslant i$) would be independent of the stock on previous days and so satisfy the Markov property (1). But the stock is increased by deliveries from orders placed several days before, and whether an order is placed will usually depend on the stock held on the day the order is placed. Hence a transition from E_i to $E_j (j > i)$ may well depend on the states assumed before E_i and the Markov property is violated.

The structure of Markov chains

As explained in §4.4 recurrent events form the natural generalization of independent trials and similarly Markov chains form the natural generalization of independent (discrete) random variables. They form a stochastic process in the sense of §4.1 with T the set of non-negative integers. The temporally homogeneous condition is also an obvious extension from the case of identically distributed random variables: the transition distributions are all identical. It is only in the case of the stationary distribution that the actual distributions over the states do not change with time. It is often said that equation (2) only obtains for Markov chains. This is not so, it is merely a consequence of the generalized addition law and is valid without equation (1). It is only useful for Markov chains since it is usually only then that $\mathbf{P}$, the matrix of elements $p(x_n = E_j \mid x_{n-1} = E_i)$, will be known: for a general process $\mathbf{P}$ will be difficult to obtain; for a Markov process $\mathbf{P}$ and $\mathbf{p}_0$ completely define the process.

Theorem 1 describes the structure of Markov chains. There are three basic types: (i) all states are transient (by theorem 3 this cannot occur with finite chains), (ii) all states are null, (iii) all states are ergodic. In case (i) each state is visited only a finite number of times and the chain is continually moving into new states—an example is the unrestricted simple random walk with $p \neq q$. In case (ii) each state occurs infinitely often but infinitely rarely and the chain never reaches a stable situation—an example is the same process with $p = q$, the variance of the distance from the origin increasing steadily. In case (iii) each state occurs infinitely often and has a finite mean recurrence time. Whatever be the initial distribution, the chain settles down to a stationary distribution and itself becomes *stationary*, neither $\mathbf{p}$ nor $\mathbf{P}$ depending on time—an example is the queueing process above with $\rho < 1$. In practice (iii) is most easily distinguished from (i) and (ii) by theorem 2 and its corollary. That is, by the stationary distribution which exists only in that case. This method was used in the queueing process. We shall see how to distinguish (i) and (ii) in the next section. From these three basic types others can be made; the fundamental, or canonical, structure being a number of closed sets of persistent states, each of which is either null or ergodic, plus transient states: once in one of the closed sets it cannot leave it. An example is the simple random walk with two absorbing barriers: there are two closed sets each of one state, E_0 and E_N, and the remaining states are transient. Examples of more than one closed set with no transient states may be ignored since they are equivalent to several separate chains (for each closed set). Notice, in the proof of theorem 2, that the basic renewal theorem 4.4.2 establishes not only the existence of the limit but its value μ_i^{-1}; although, in practice, μ_i is often most easily found as p_i^{-1}, p_i being found from (9).

Finite Markov chains

Finite Markov chains have some special properties, principally that null states cannot occur and transient states can only occur at the beginning of the process. Interest therefore centres on the time taken to leave the transient states (to be discussed in the next section) and on the stationary distribution. The computation of the latter is particularly interesting. Let us consider a finite aperiodic chain with a single closed set and no transient states. We have to solve the equation in $\mathbf{z}$

$$\mathbf{z}' = \mathbf{z}'\mathbf{P}, \tag{15}$$

knowing it to have a unique solution with $\sum_i z_i = 1$, and therefore, if $\mathbf{y}$ is any other solution $\mathbf{y} = \mathbf{z}(\sum_i y_i)$. Writing (15) in the form $\mathbf{z}'(\mathbf{I}-\mathbf{P}) = \mathbf{0}$, where $\mathbf{I}$ is the unit matrix, we see that our results imply that 1 is a characteristic root of $\mathbf{P}$ (since $\mathbf{z}$ exists) and is a simple root (since $\mathbf{z}$ is unique). This suggests we consider the other characteristic roots, λ_s, and characteristic vectors, $\mathbf{z}_s$, satisfying $\mathbf{z}_s'(\lambda_s\mathbf{I}-\mathbf{P}) = \mathbf{0}$ with $\mathbf{z}_1 = \mathbf{z}$ corresponding to $\lambda_1 = 1$. Now it is a known result in the theory of such roots and vectors that the $\mathbf{z}_s$ span the N-dimensional space of distributions: that is, that any $\mathbf{p}_0$ can be written as a linear combination of the $\mathbf{z}$'s;

$$\mathbf{p}_0' = \sum_s a_s \mathbf{z}_s'$$

for suitable a_s. It follows that

$$\mathbf{p}_1' = \mathbf{p}_0'\mathbf{P} = \sum_s a_s \mathbf{z}_s'\mathbf{P} = \sum_s a_s \lambda_s \mathbf{z}_s'$$

and generally

$$\mathbf{p}_n' = \sum_s a_s \lambda_s^n \mathbf{z}_s'. \tag{16}$$

Since we know $\lim_{n\to\infty} \mathbf{p}_n' = \mathbf{z}' = \mathbf{z}_1'$, we must have $a_1 = 1$ and $|\lambda_s| < 1$ for all $s > 1$. Equation (16) is a convenient form for calculating $\mathbf{p}_n$ involving only the calculation of the a's, $\mathbf{z}$'s and λ's and then raising the last to the appropriate powers. Moreover, (16) tells us how rapidly the limit is approached because this will depend primarily on the root of second largest modulus.

If the chain is periodic then there will exist roots with $|\lambda_s| = 1$ other than $\lambda_1 = 1$. Thus if it has period 2 there will be a root, say λ_2, with $\lambda_2 = -1$; and considering the chain only after an even number of transitions we shall have

$$\lim_{n\to\infty} p_n' = a_1 \mathbf{z}_1' + a_2 \mathbf{z}_2'$$

since $(\lambda_2)^{2n} = 1$. If the chain also has transient states then the limiting distribution over all states will be the limiting distribution over the states of the closed set with zero added for the transient states. If there are several closed sets there will be a limiting distribution for each set and $\lambda = 1$ will have multiplicity equal to the number of sets. It is instructive to note that we here proved purely algebraic results concerning the roots of $\mathbf{P}$, by probabilistic methods.

Applications to genetics

The genetical example of inbreeding considered in §1.5 is closely related to a Markov chain. In the genetical situation we were considering the proportions of a population in three states (*AA*, *Aa* and *aa*). Remembering the intimate relationship, discussed in §1.3, between probability and population proportions, it is easy to see that the inbreeding example is a Markov chain with three states *AA*, *Aa* and *aa*. The Markov property is expressed in the fact that given the parents' genetical structure (generation r, say) the offsprings' genetical structure (at the $(r+1)$st generation) is independent of all the grandparents, greatgrandparents, ... genetical structure (at generations previous to the rth). The chain described by equation 1.5.5 had two closed sets, AA and aa, and one transient state Aa. Hence the matrix had the root $\lambda = 1$ with multiplicity two. The single remaining root expresses the rapidity with which the chain leaves the transient state.

Miscellaneous remarks

It is possible to extend Markovian concepts to continuous time processes (such as those of §§4.1 and 4.2) and to continuous variables. Indeed the queueing process of §4.2 in continuous time has the Markov property, whereas that of §4.3 does not and, as explained in §4.3, this is the reason for considering instead the Markov chain at points where customers leave the system. The Markovian property is basic to the study of most stochastic processes.

The results of this section enable us to distinguish ergodic states from others by considering the existence of a stationary distribution. We have still to find how to distinguish null and transient states and to find how long a chain stays in the latter. These problems are studied in the next section.

4.6. Markov chains (continued)

The study begun in the previous section is continued with the same notation.

Theorem 1. *Let E_i be a transient state and denote by π_i the probability that, starting from $x_0 = E_i$, the chain will forever stay in transient states. Then the π_i satisfy*

$$\pi_i = \sum_T p_{ij}\pi_j, \tag{1}$$

where the summation is over all j such that E_j is transient. Furthermore $\{\pi_i\}$ is the largest solution of (1) *bounded by* 1.

It is obvious that the π_i satisfy (1). For consider the transition from x_0 $(= E_i)$ to x_1. The only way the chain can remain in transient states for ever is to pass to a transient E_j, an event of probability p_{ij}, and then, starting in E_j, remain in transient states forever. But by the Markov property this latter probability is π_j. The possibilities, for different j, are exclusive and exhaustive, hence (1).

It remains to prove the final sentence. Let z_i be any solution of (1) bounded by 1: that is with $|z_i| \leqslant 1$. Let $\pi_i^{(n)}$ denote the probability that, starting from $x_0 = E_i$, the chain will still be in transient states after n transitions. Then, exactly as (1),

$$\pi_i^{(1)} = \sum_T p_{ij}, \tag{2}$$

$$\pi_i^{(n)} = \sum_T p_{ij}\pi_j^{(n-1)}. \tag{3}$$

Also
$$z_i = \sum_T p_{ij} z_j. \tag{4}$$

A comparison of (4) with (2) shows that, since $|z_j| \leqslant 1$, $|z_i| \leqslant \pi_i^{(1)}$ and then by induction, using (3), $|z_i| \leqslant \pi_i^{(n)}$ for all n. But since the event of being in transient states after n transitions implies being in transient states after $(n-1)$, $\pi_i^{(n)} \leqslant \pi_i^{(n-1)}$ (theorem 1.4.3). Hence $\pi_i^{(n)}$ tends to a limit which is clearly π_i and $|z_i| \leqslant \pi_i$. This proves the result.

Corollary. *If (1) has no solutions bounded by 1 other than $z_i = 0$ then the chain cannot stay in transient states for ever.*

This is immediate because $z_i = 0$ is the largest solution bounded by 1.

Consider a chain with closed sets and transient states. Let ρ_{ij} be the probability that, starting from a transient state E_i, the chain will at some time be in the persistent state E_j. Now if it reaches E_j it is certain (theorem 4.5.1) to reach all states in the closed set to which E_j belongs. Consequently ρ_{ij} depends, not on j, but on i and the closed set. Consider, for the moment, a particular closed set C and let ρ_i denote the probability, starting from E_i, that the chain will eventually enter C (and therefore stay in C).

Theorem 2. *The ρ_i satisfy*

$$\rho_i = \sum_T p_{ij}\rho_j + \sum_C p_{ij}, \tag{5}$$

where the second summation is over all states E_j in C. Furthermore $\{\rho_i\}$ is the smallest positive solution of (5).

Consider the transition from x_0 ($=E_i$) to x_1. It can either pass into C at x_1, with probability $\sum_C p_{ij}$, or can pass to another transient state and pass from there into C. As in theorem 1 this probability is $\sum_T p_{ij}\rho_j$. Hence (5) follows.

The proof of the final sentence proceeds parallel to the proof in theorem 1. If $\rho_i^{(n)}$ is the probability, starting from E_i, of reaching C in n transitions or less, then

$$\rho_i^{(1)} = \sum_C p_{ij}, \tag{6}$$

and

$$\rho_i^{(n)} = \sum_T p_{ij}\rho_j^{(n-1)} + \sum_C p_{ij}. \tag{7}$$

If z_i a positive solution of (5) we clearly have $z_i \geqslant \sum_C p_{ij} = \rho_i^{(1)}$ and hence, by induction using (7), $z_i \geqslant \rho_i^{(n)}$. But $\rho_i^{(n)} \geqslant \rho_i^{(n-1)}$ and $\lim_{n\to\infty} \rho_i^{(n)} = \rho_i$, whence $z_i \geqslant \rho_i$.

Note that if there is only one closed set, (5) immediately follows from (1) since $\pi_i + \rho_i = 1$. Also, any two solutions of (5) differ by a solution of (1).

A chain in which every state can be reached from every other is *irreducible*. From theorem 4.5.1 the states of an irreducible chain must either all be transient, null or ergodic. We have seen how to distinguish the last case, by the existence of a stationary distribution (theorem 4.5.2). The above results can be used to distinguish the first case.

Theorem 3. *A necessary and sufficient condition that all the states of an irreducible chain be transient is that the equations*

$$z_i = \sum_{j>0} p_{ij} z_j \tag{8}$$

admit of a non-zero bounded solution. (Note the range of summation.)

The proof uses a commonly useful device. Change the chain by making any single state, say E_0, absorbing. That is, alter p_{0j} so that $p_{00} = 1$, $p_{0j} = 0$ ($j > 0$). Then if the states of the original chain were persistent one would be certain (theorem 4.5.1) to pass into E_0 and therefore in the new chain absorption in E_0 would be certain and the other states would all be transient. Conversely if, in the new chain, absorption in E_0 is certain one must, in the original chain, have been certain to reach E_0 from any other state and therefore E_0, and hence all other states, would have been persistent. Hence a necessary and sufficient condition that all states of the original chain be transient is that it is possible, in the new chain, to stay for ever in transient states (that is, states other than E_0). But the necessary and sufficient condition for the latter is that (1) has a non-zero bounded solution. In our case (1) reduces to (8) and the theorem is proved.

Theorem 4. *In a chain containing one or several closed sets and transient states, in which it is impossible to stay in transient states for ever, let ν_i be the expected number of transitions, starting in a transient state E_i, before first entering any closed set. Then $\{\nu_i\}$ satisfies the equations*

$$\nu_i = \sum_T p_{ij}\nu_j + 1. \tag{9}$$

Furthermore, ν_i is the minimal positive solution of (9).

Let $\tau_i^{(n)}$ be the probability, starting from E_i, that the chain reaches a closed set for the first time on the nth transition. Then $\sum_n \tau_i^{(n)} = 1$ and $\sum_n n\tau_i^{(n)} = \nu_i$. ($\tau_i^{(n)}$ should be carefully distinguished from $\rho_i^{(n)}$ in equations (6) and (7).) Clearly, using arguments similar to those used above, if C denotes the union of all closed sets,

$$\left.\begin{aligned} &\tau_i^{(1)} = \sum_C p_{ij}, \\ \text{and, for } n > 1, \quad &\tau_i^{(n)} = \sum_T p_{ij}\tau_j^{(n-1)}. \end{aligned}\right\} \tag{10}$$

Multiply the nth equation by n and sum. We obtain

$$\begin{aligned} \nu_i &= \sum_n \sum_T p_{ij}(n-1)\,\tau_j^{(n-1)} + \sum_n \sum_T p_{ij}\tau_j^{(n-1)} + \sum_C p_{ij} \\ &= \sum_T p_{ij}\nu_j + \sum_T p_{ij} + \sum_C p_{ij} \\ &= \sum_T p_{ij}\nu_j + 1 \end{aligned}$$

as required.

To prove that we require the minimal solution of (9) we consider

$$\nu_i^{(n)} = \sum_{r=1}^{n} r\tau_i^{(r)}.$$

As in the proof of a similar result in theorem 1 we easily establish that if z_i is any positive solution of (9) then $\nu_i^{(1)} \leqslant z_i$ and that if $\nu_i^{(n)} \leqslant z_i$ so is $\nu_i^{(n+1)} \leqslant z_i$. Consequently $\nu_i^{(n)} \leqslant z_i$ for all i. But $\nu_i^{(n)} \leqslant \nu_i^{(n+1)}$, $\nu_i = \lim_{n\to\infty} \nu_i^{(n)}$ and the result is established.

Structure of Markov chains

In the previous section it was explained that there exist three basic types of Markov chain: in which all states are either (i) transient, (ii) null, (iii) ergodic. We saw that case (iii) could be distinguished by the existence of a stationary distribution. Theorem 3 of this section enables case (i) to be recognized and hence all three basic types can be separated. Other chains can be built up from these. Theorems 1 and 2 enable a chain with transient and persistent states to be studied by calculating the probabilities of for ever remaining in transient states and of passing to persistent states: once it is in the persistent states the stationary distribution applies if the persistent states are ergodic. Note that in such a chain, if E_i is transient and E_j persistent,

$$\lim_{n\to\infty} p_{ij}^{(n)} = \rho_i \mu_j^{-1},$$

since ρ_i is the probability of reaching the closed set containing E_j and μ_j^{-1} is the conditional probability of being in E_j given that the chain is in the closed set.

Examples

These results can be applied to the examples of the previous section. Examples (*a*) and (*c*) are completely solved.

(*b*) *Two absorbing barriers.* We saw that absorption in E_0 or E_N was certain. Let ρ_i denote the probability, starting from E_i, of reaching the closed set containing the single state E_0. Then equation (5) gives, for $0 < i < N$,

$$\rho_i = p\rho_{i+1} + q\rho_{i-1} \tag{11}$$

with $\rho_0 = 1$, $\rho_N = 0$. This equation is the same as equation 4.5.13 for $0 < i < N$ but with p and q interchanged. The general solution is therefore

$$\rho_i = A(q/p)^i + B$$

with boundary conditions $\rho_0 = 1$, $\rho_N = 0$. Determination of A and B to satisfy these conditions gives easily

$$\rho_i = \left\{\left(\frac{q}{p}\right)^i - \left(\frac{q}{p}\right)^N\right\} \Big/ \left\{1 - \left(\frac{q}{p}\right)^N\right\}. \tag{12}$$

This solution fails if $p = q$. A special argument for this case, or allowing p/q to tend to 1 in (12), gives

$$\rho_i = (N-i)/N. \tag{13}$$

Naturally both (12) and (13) diminish as i increases: that is, the farther the chain starts from E_0 the less likely it is to reach it. The probability of absorption in E_N is clearly $1-\rho_i$, or alternatively is the solution of (11) with $\rho_0 = 0$, $\rho_N = 1$.

Theorem 4 enables the expected number of transitions before absorption in either boundary to be found. Equations (9) give for $0 < i < N$

$$\nu_i = p\nu_{i+1} + q\nu_{i-1} + 1 \tag{14}$$

with $\nu_0 = \nu_N = 0$. These equations differ from (11) only in the extra unit term on the right. The general solution of (14) is therefore the general solution of (11) plus a particular solution of (14). An example of the latter, not included in the former, is $i/(q-p)$ so that the general solution of (14) is

$$A(q/p)^i + B + i/(q-p).$$

The boundary conditions $\nu_0 = \nu_N = 0$ determine A and B and the final result is

$$\nu_i = \frac{N}{p-q}\left\{\frac{(q/p)^i - 1}{(q/p)^N - 1}\right\} + \frac{i}{q-p}. \tag{15}$$

This fails if $p = q$ when a limiting procedure or a fresh solution of the equations gives the result

$$\nu_i = i(N-i). \tag{16}$$

These results are relevant to the game of chance interpretation of the random walk mentioned in §2.5. If player A has an initial capital of i units and B has $(N-i)$ units: and if the probability of A winning a single play is p, then the chance of B winning (that is, A's capital reducing to zero) is given by ρ_i (equations (12) or (13)) and the expected duration of the game is given by ν_i (equations (15) or (16)).

(*d*) *One absorbing barrier.* We saw in the last section that E_0 is absorbing and E_i ($i \geqslant 1$) transient. But it remains to see whether absorption is certain or whether it can stay forever in transient states. The corollary to theorem 1 enables this to be decided. We have to consider the equations, for $i > 0$,

$$\pi_i = p\pi_{i+1} + q\pi_{i-1} \tag{17}$$

with $\pi_0 = 0$. These are (11) again, and have, using the condition $\pi_0 = 0$, the general solution $A[(q/p)^i - 1]$. They have a bounded non-zero solution iff $q < p$. Thus absorption is certain if $q \geqslant p$. If $q < p$ then the probability of *not* being absorbed is the maximal solution of (17), bounded by 1, which is clearly seen to be

$$\pi_i = 1 - (q/p)^i. \tag{18}$$

The probability of absorption is therefore 1 if $q \geqslant p$ and $(q/p)^i$ if $q < p$. In the former case the expected time to absorption can be found from theorem 4. The equations to be solved are, for $i > 0$,

$$\nu_i = p\nu_{i+1} + q\nu_{i-1} + 1 \tag{19}$$

with $\nu_0 = 0$. This is (14) again, and has, using the condition $\nu_0 = 0$, the general solution $A[(q/p)^i - 1] + i/(q-p)$. The minimal positive solution of this has $A = 0$, since $A < 0$ would make ν_i negative for sufficiently large i; consequently

$$\nu_i = i/(q-p). \tag{20}$$

If $p = q$ the expected time to absorption is infinite.

(*e*) *One impenetrable barrier.* If $p < q$ we found a stationary distribution. It remains to consider $p \geqslant q$, the only possibilities are that all states are null or all transient. This is settled using theorem 3. The proof of that theorem involves treating E_0 as absorbing; that is, replacing it by case (*d*) just studied. The equations (8) are then (17) and admit of a bounded non-zero solution in the cases $p \geqslant q$ only if $p > q$. Consequently if $p > q$ all states are transient, if $p = q$ all states are null.

(*f*) *Queueing theory.* This is closely similar to (*e*). We leave it as an exercise for the reader to show that with $\rho = 1$ all states are null but with $\rho > 1$ they are transient. Thus if the traffic intensity is one the queue will be certain to return to zero size but will do so with infinite mean recurrence time; that is, infinitely rarely.

Forward and backward equations

These examples exhibit the main uses of the theorems. It is worth noticing some differences in character between the equations of this section and those of previous sections. Let us contrast, for example, equations 4.5.2 and (1). In the former we are considering 'now' after the nth transition and considering how it can have arisen from developments in the 'past', that is after the $(n-1)$st transition. We are, as we have done so often with the other processes studied in this chapter, looking *back* at the development of the process. In (1), by contrast, we consider 'now' in relation to the 'future', by relating x_0 to x_1: we see how it develops *forward* in time. Which way one has to look depends on what is wanted in the process. Both approaches lead to useful results.

A further point to notice in connexion with the methods used in this section is that they all solve more than is normally asked for. For example, suppose, in the game of chance mentioned in connexion with example (*b*) above, we are interested in ρ_i or ν_i. In order to find these we find also the probabilities and expectations for other values of i. This technique basically consists in relating the value for one i with the values for other suffixes.

Method of change of state

The method used, in proving theorem 3, of changing a state into an absorbing state and thereby producing a closed set containing this one state is often useful. Consider, for example, a chain containing a single closed set of ergodic states. We know how long it will take on the average to return to a given state, namely the mean recurrence time of that state. A related quantity is how long it will take on the average to pass from E_i to a different state E_j. We know it is certain to happen but how long will it take? This is easily solved by making E_j absorbing but otherwise leaving the chain unaltered. All states, except E_j, will now be transient and the quantity required is the expected time to absorption. This is easily found by theorem 4, at the cost of also finding the time to pass from any E_i to E_j. If we rewrite μ_i as μ_{ii} then we have, by this device, found μ_{ij}. More complicated problems can similarly be solved this way. For example, we may wish to know the probability, starting from E_i, of reaching E_j before E_k (i, j, k all different). This is solved by making E_j and E_k absorbing and calculating (much as in example (*b*) above) the probability of absorption in E_j. This is clearly equal to the desired probability.

Suggestions for further reading

There is an extensive literature on queues. Much of the earlier work is given in the collected works of Erlang, edited by Brockmeyer *et alii* (1948). A good introduction is provided by Khintchine (1960) and Cox and Smith (1961). There is an excellent account of Markov chains in Feller (1957). An elementary introduction to finite Markov chains is provided by Kemeny and Snell (1960). More advanced books on this subject are by Chung (1960) and Dynkin (1960). An account of Renewal Theory is given by Cox (1962).

A wider field of stochastic processes is covered in the books by Doob (1953) and Bartlett (1956). In the former the emphasis is on the mathematics, whilst the latter provides a treatment balanced between the mathematics and its applications.

Exercises

1. Construct some realizations of immigration–emigration processes using random number tables as in fig. 4.1.1. Do the same for a queueing process with Poisson arrivals and a Γ-distribution of index 2 for the service-time, plotting the number in the queue and the waiting times.

2. The usual immigration–emigration process is modified so that a particle arriving and finding N particles in the region does not enter it but leaves and is 'lost'. Find the stationary distribution of the number of particles in the region.

3. Show that in an immigration–emigration process the distribution of the number of particles in the region at time t which were not in it at time zero is Poisson with mean

$$\lambda(1-e^{-\kappa t})/\kappa.$$

4. In an immigration–emigration process in one dimension the region is an interval of unit length. The particles all enter at the left-hand end and leave at the right, travelling at constant speed. Find the distribution of speed of the particles and show that its mean is infinite.

The process described above is observed at a random time and the speed of a randomly chosen particle is noted. Ignoring the possibility of the interval being empty (λ/κ large) show that this particle has a distribution of speed v, equal to

$$\kappa^2 v^{-3} e^{-\kappa/v}.$$

Show that this has finite mean. Discuss the difference between this result and that in the first paragraph.

5. A certain type of bacterium grows in a colony in such a way that the probability that any one individual gives birth to a new individual in any time interval of length δt is $\lambda\delta t + o(\delta t)$ independent of the past history of the individual or of other individuals. Obtain a differential equation for the probability, $p_n(t)$, that there will be n individuals in the colony at time t, given that there was a single individual at $t = 0$. Show that

$$p_n(t) = e^{-\lambda t}(1-e^{-\lambda t})^{n-1} \quad (n \geqslant 1).$$

The colony grows for a time t in the way just explained. Growth then stops and any individual then has probability $\mu\delta t + o(\delta t)$ of dying in an interval δt independently of its past history or of the other individuals. Show that after a total time $t+s$ (t spent growing, s spent decaying) the probability that the original individual has no descendants left is

$$(e^{\mu s}-1)/(e^{\lambda t}+e^{\mu s}-1).$$

(Wales Dip.)

6. In a simple birth process each individual has a constant chance $\lambda\delta t + o(\delta t)$ of giving birth to a new individual in $(t, t+\delta t)$, independently of the previous history of the individual. Events for different individuals are independent. At $t = 0$ there is one individual.

Let T_m be the time at which the population size first reaches the value m, where $m > 1$ is a given integer. Prove that the density of T_m is

$$\lambda(m-1)(1-e^{-\lambda t})^{m-2} e^{-\lambda t}.$$

(Lond. M.Sc.)

7. In a simple model for the growth of a bacterial population it is assumed that the probability of a bacterium dividing to give two new bacteria during the time interval $(t, t+\delta t)$ is $\lambda\delta t + o(\delta t)$, and the probability of more than one division during $(t, t+\delta t)$ is $o(\delta t)$, these probabilities for different bacteria all being mutually independent.

At each division of a normal bacterium there is a probability p that *one* of the two bacteria produced is a mutant and a probability $(1-p)$ that both are normal, and that a division of a mutant bacterium must produce two mutants, the probabilities for different divisions being mutually independent. Obtain the 'forward' partial differential equation satisfied by the probability generating function of the joint distribution of $N_1(t)$, the number of normal, and $N_2(t)$, the number of mutant, bacteria. Hence, or otherwise, determine $\mathscr{E}\{N_2(t)\}$, and show how expressions for $\mathscr{C}\{N_1(t), N_2(t)\}$ and $\mathscr{D}^2\{N_2(t)\}$ may be found, given that $N_1(0) = n, N_2(0) = 0$.

(Camb. Dip.)

8. In a density-dependent population of unicellular organisms the number of individuals varies between 1 and N. If at time t there are n in the population the chance that one individual will die in a small interval of time δt is $\mu(n-1)\delta t$, whereas the chance that one will divide into two similar individuals is $\lambda(N-n)\delta t$. Obtain the equilibrium distribution of n and deduce that for N and t large

$$N^{\frac{1}{2}}\left\{\frac{n}{N}-\frac{\lambda}{\lambda+\mu}\right\}$$

in approximately normally distributed with zero mean and finite variance.

(Lond. Dip.)

9. Events occur in a Poisson process of rate λ. With the ith event is associated a random variable X_i having density $\rho e^{-\rho x}$ $(x > 0)$. The random variable $Y(t)$ is defined as the sum of the X_i over all events occurring in $(0, t)$; $Y(t)$ is zero if no events have occurred by time t. Prove that $p\{Y(t) = 0\} = e^{-\lambda t}$ and that the density of $Y(t)$ is, for non-zero y,

$$\frac{1}{2}\left(\frac{\rho\lambda t}{y}\right)^{\frac{1}{2}} e^{-(\rho y+\lambda t)} I_1\{2(\lambda t\rho y)^{\frac{1}{2}}\},$$

where $I_1(u)$ is the first-order Bessel function of imaginary argument,

$$I_1(u) = \sum_{r=0}^{\infty} \frac{(\frac{1}{2}u)^{2r+1}}{r!(r+1)!}.$$

The random variable T is defined as the time at which the process $\{Y(t)\}$ first reaches or crosses a barrier at $y = a$. Obtain the density of T.

(Lond. M.Sc.)

10. Events occur randomly in time in a Poisson process with rate of occurrence λ. The random variable X is independent of the Poisson process and has distribution function $F(x)$. The number of events that occur in the time interval $(0, X)$ is denoted by Y. Obtain the probability generating function of Y in terms of the moment generating function of X.

Suppose now that there are two independent Poisson processes. One is the process of events referred to above, and the other is a process of catastrophes occurring with rate of occurrence μ. The random variable X is defined to be the time of the occurrence of the first catastrophe, so that Y is the number of events that occur before there is a catastrophe. Prove that

$$p(Y=r) = \frac{\mu\lambda^r}{(\mu+\lambda)^{r+1}}. \qquad \text{(Lond. M.Sc.)}$$

11. Carry through the analysis of §4.3 in the special case where each customer takes the same time, κ^{-1} (to conform to earlier notation), to be served.

12. Consider a simple queueing process with the modification that the customers arrive in groups of 2 but are served individually as before. (The groups now arrive at rate λ.) Determine

(i) the stationary distribution of queue size, and the conditions under which it exists:

(ii) the distribution of queueing time, assuming that the customers are served in their group order, but at random within a group.

13. You are in a queue where the successive service-times are independently and exponentially distributed. Show that the probability that you will have to wait for more than 3 times as long as the person in front of you is 1/4.

14. In a simple queueing process, show that the probability generating function

$$\phi(s, t) = \sum_{n=0}^{\infty} p_n(t) s^n$$

satisfies the differential equation

$$\frac{\partial \phi(s, t)}{\partial t} = \frac{s-1}{s} [(\lambda s - \kappa)\phi(s, t) + K p_0(t)].$$

Deduce that if the mean number of individuals in the queue is always unity, the entire probability distribution, $p_n(t)$, is independent of time with a variance equal to 2. (Camb. Trip.)

15. Consider the simple queueing system. Assuming the traffic intensity to be less than 1 and the system to have been in operation for a sufficiently long time for equilibrium probabilities to hold, find the probability distribution of the time interval between a departure of a randomly selected customer and the immediately following departure of a customer. (Aberdeen Dip.)

16. In a simple queueing process the probability that a customer joins the queue in $(t, t+\delta t)$ is $\lambda_n \delta t$, independent of the behaviour of the system prior to t, where n is the size of queue at t. Similarly the probability that a customer leaves is $\kappa_n \delta t$. Show that an equilibrium distribution only exists if

$$1 + \frac{\lambda_0}{\kappa_1} + \frac{\lambda_0 \lambda_1}{\kappa_1 \kappa_2} + \frac{\lambda_0 \lambda_1 \lambda_2}{\kappa_1 \kappa_2 \kappa_3} + \dots$$

converges. Consider the special cases:

(i) $\lambda_n = \lambda/(n+1)$, $\kappa_n = \kappa$ (a long queue discourages new arrivals). Show that the stationary queue size distribution is now Poisson.

(ii) $\lambda_n = \lambda$, $\kappa_n = n\kappa$ for $n \leqslant m$, $\kappa_n = m\kappa$ for $n \geqslant m$. (There are m servers attending to the customers.) Obtain the stationary distribution of queue size.

17. A queueing process has two stages of service. Customers arrive in a Poisson process at rate α and queue for service at the first stage. There is one server and the distribution of service-time is exponential with mean $1/\sigma_1$. A customer who has been served at the first stage passes immediately to the second stage where there is a single server and where the distribution of service-time is exponential with mean $1/\sigma_2$. The stochastic processes

$N_1(t)$, $N_2(t)$ are the numbers of customers in the first and second stages of the system at time t; $N_1(t)$ is thus the number of customers queueing and being served in the first stage.

Show that in statistical equilibrium N_1 and N_2 are distributed in independent geometric distributions. Write down the condition for the existence of an equilibrium distribution, and explain the practical interpretation of the equilibrium distribution. (Lond. M.Sc.)

18. In a certain queue, it takes a constant time T to serve any customer. A_m is the probability that m customers arrive in any period T, and P_m is the probability that exactly m customers are in the queue in any period T. Show that

$$P_m = A_m P_0 + A_m P_1 + A_{m-1} P_2 + \ldots + A_0 P_{m+1},$$

and deduce that, if

$$a(x) = \sum_0^\infty A_m x^m,$$

$$f(x) = \sum_0^\infty P_m x^m,$$

then

$$f(x) = \frac{a(x)(1-x)}{a(x)-x} f(0).$$

Show that the average number of customers waiting to be served at any time is

$$E = \frac{\frac{1}{2}\eta}{1-\epsilon},$$

where $\epsilon = a'(1)$ and $\eta = a''(1)$. (Camb. Trip.)

19. A bus travels along a road of length a. In the course of its journey the bus is subject to accidental delays; it will be supposed that the total delay at any stage of the journey is an integral number of minutes. The delays are subject to the law that in any small segment $(x, x+\Delta x)$ of the road there is a probability $\lambda\Delta x + o(\Delta x)$ of the occurrence of a delay of 1 minute, where λ is a constant. Show that the total delay at the end of the journey has a probability distribution following a Poisson law with mean value λa.

A second bus starts on the same journey m minutes after the first one. Show that the probability q_n that the second bus arrives n minutes later than the first one is equal to the coefficient of s^{n-m} in the Laurent expansion of

$$\exp[\lambda a(s-2+s^{-1})],$$

and deduce that

$$q_n = \frac{1}{2\pi}\int_{-\pi}^{\pi} \exp[-2\lambda a(1-\cos\theta)]\cos(n-m)\theta\, d\theta.$$

(Camb. Trip.)

20. A stochastic process consisting of a series of trials can be in either of two states A or B. If at trial n it is in state A and was last in state B at trial $n-r$ then the probability that at trial $(n+1)$ it will be in state B is λ_r $(r = 1, 2, \ldots)$ independent of all states previous to the $(n-r)$th trial. The same probability with A and B interchanged is μ_r. The process starts at the first trial in state A with a fictitious zero trial in which it was in state B. Identify two recurrent events. Obtain a necessary and sufficient condition for (i) B to be certain to occur, and (ii) for A to be certain to occur again after B has occurred. Under these conditions find the limiting probability that the process will be in state A.

21. A series of trials is made in each of which an event A is observed to occur or not. The probability that it will occur at any trial is p_s, where s is the number of trials since A last occurred ($s = 1, 2, \ldots$), and is independent of all events other than the last occurrence of A.

(i) If $p_s = (s+1)^{-1}$ show that once A has occurred it is certain to occur again, but that the mean time between occurrences is infinite.

(ii) If $p_s = 1-s^{-1}$ show that the mean and variance of the time between occurrences are e and $e(3-e)$ respectively. (Camb. Trip.)

22. A stochastic process in continuous time is specified by the following transition probabilities between its three states in the small time interval $(t, t+\delta t)$, the remainder terms of $o(\delta t)$ being omitted for convenience and all being independent of the process prior to t:

(i) in state 1 at t, equal probabilities δt of passing to states 2 and 3;

(ii) in state 2 at t, probability δt of passing to state 1, probability $2\delta t$ of passing to state 3;

(iii) in state 3 at t, probability $3\delta t$ of passing to state 1, probability δt of passing to state 2.

The recurrence time T of any state is defined as the time from leaving that state to first returning to it again. If the distribution of T is defined by the density $f(T)\,dT$, and

$$L(\psi) = \int_0^\infty e^{-\psi T} f(T)\,dT,$$

obtain $L(\psi)$ for the recurrence time of state 1. If, alternatively, T is defined as including the time before leaving state 1 (given that it is occupied at $t = 0$), as well as the subsequent time before returning to it, what would $L(\psi)$ be then? (Lond. Dip.)

23. The independent random variables $X_1, X_2, \ldots$ take the values $-1, +1, +2$, each with probability $1/3$. Show that

$$p_m = (2^{\frac{1}{2}}-1)^m,$$

where m is a positive integer and p_m is the probability that

$$X_1 + \ldots + X_n = -m$$

for some n. (Camb. Trip.)

24. The life-time up to failure of electric light bulbs of a certain type is a non-negative random variable having probability density function $f(x)$. The probability of a failure in $(x, x+\delta x)$ given that no failure has occurred up to x is $\lambda(x)\delta x + o(\delta x)$.

Express $\lambda(x)$ in terms of $f(x)$, and hence show that

$$f(x) = \lambda(x)\exp\left\{-\int_0^x \lambda(u)\,du\right\}. \qquad \text{(Lond. B.Sc.)}$$

25. In a renewal process events occur at times $X_1, X_1+X_2, X_1+X_2+X_3, \ldots$, where $\{X_i\}$ are a set of mutually independent identically distributed random variables, with density $f(x)$. Let N_t be the number of renewals in $(0, t)$ and let $H(t) = \mathscr{E}(N_t)$. Show that

$$H^*(s) = \frac{f^*(s)}{s[1-f^*(s)]},$$

where the asterisk denotes a Laplace transform, for example

$$f^*(s) = \int_0^\infty e^{-st} f(t)\,dt.$$

Hence show that

$$H(t) - \frac{t}{\mu} \sim \frac{\mu^2-\sigma^2}{2\mu^2} \quad \text{as} \quad t \to \infty,$$

where μ and σ are the mean and standard deviation of the distribution $f(x)$.
Compare the exact and asymptotic values of $H(t)$ when

$$f(x) = \tfrac{1}{2}e^{-x} + e^{-2x}. \qquad \text{(Lond. M.Sc.)}$$

26. A signalling apparatus can either be 'on' or 'off'. If it is 'off' at time t the probability that it will come 'on' in the interval $(t, t+\delta t)$ is $\lambda\delta t + o(\delta t)$. If it is 'on' at time t the probability that it will come 'off' in the interval $(t, t+\delta t)$ is $(\mu\delta t + o(\delta t))$. These probabilities are independent of the behaviour of the system prior to t. Obtain a differential equation for $p(t)$, the probability that at time t, the signal is 'off'. If it is 'off' at time $t = 0$ show that

$$p(t) = \{\lambda e^{-(\lambda+\mu)t} + \mu\}/(\lambda+\mu)$$

and hence obtain $\lim_{t\to\infty} p(t)$.
Obtain this last limiting result by an ergodic argument.

(Wales Maths.)

27. The random variables $X_1, X_2, X_3, \ldots$, may each take only the values 0 or 1; their joint probability distribution is defined by the relations

$$P(X_1 = 1) = \tfrac{1}{2},$$

$$P(X_{n+1} = X_n \mid X_1 = x_1, X_2 = x_2, \ldots, X_n = x_n) = 1 - \alpha_n,$$

for all $(x_1, x_2, \ldots, x_n)$ and $n = 1, 2, 3, \ldots$. The sequence of random variables $\{Y_n\}$ is defined by

$$Y_n = \frac{1}{n}\sum_{i=1}^{n} X_i.$$

Establish whether the property '$\{Y_n\}$ converges in probability to $\tfrac{1}{2}$ as $n \to \infty$' holds in the following two cases:

(i) if all $\alpha_n = \tfrac{1}{2}$;
(ii) if the infinite series $\Sigma\alpha_n$ converges. (Camb. Trip.)

28. A loom stops from time to time and the number X of stops in unit running time may be assumed to have a Poisson distribution of mean μ. For each stop there is a probability θ that a fault will be produced in the fabric being woven. Occurrences associated with different stops may be assumed independent. Let Y be the number of fabric faults so produced in unit running time. By first finding the conditional distribution of Y, given that $X = x$, or otherwise, prove that Y has a Poisson distribution of mean $\theta\mu$. Show also that the covariance of X and Y is $\theta\mu$.

(Lond. B.Sc.)

29. Each trial is an occasion on which a piece of equipment is used. The event E is the equipment's failure and replacement by a new piece which is then used, necessarily without failure on that occasion. All pieces are taken randomly from a large stock. F_n is the probability that a piece will last for at least n trials without failure ($n = 1, 2, \ldots; F_1 = 1$). The 'age' of a piece in any trial is the number of trials (excluding the one being considered) for which it has been used. State, without proof and in terms of $\{F_n\}$, the probability that, in a randomly selected trial, the piece will fail. Deduce that the probability that, in a randomly selected trial, the piece in use will have age s is

$$F_{s+1} \Big/ \sum_{i=1}^{\infty} F_i = g_s, \quad \text{say.}$$

Show that the probability that, in a randomly selected trial, the piece then in use will last a further s trials before the trial in which it fails is also g_s. Show that, in these conditions, the expected number of further trials before it fails is

$$(\sigma^2 + \mu^2 - \mu)/2\mu,$$

where μ and σ^2 are respectively the mean and variance of the age of a piece at failure.

(Wales Dip.)

30. In a sequence of trials E is an ergodic recurrent event. A trial a long time from the first is observed and s is the number of trials since E was last observed. (Thus $s = 0$ if E occurred at the observed trial, $s = 1$ if E did not occur at the observed trial but did occur at the immediately preceding one, and so on.) Show that s has density

$$p_s = \mu^{-1}\left(1 - \sum_{i=1}^{s} f_i\right).$$

Find the mean and variance of s in terms of the moments of the density f_i.

Similarly, s' is the number of trials until E is next observed. (Thus $s' = 0$ if E occurred at the observed trial, $s' = 1$ if E did not occur at the observed trial but did occur at the immediately succeeding one, and so on.) Find the density p'_s of s.

31. Consider a Markov process in discrete time, with states labelled by $j = 0, 1, 2, \ldots$, and with transition probabilities

$$p_{j,j+1} = p_j, \quad p_{j,0} = q_j = 1 - p_j, \quad p_{jk} = 0 \quad (k \neq 0, j+1).$$

Consider the recurrent event E: occupation of the state $j = 0$. Determine conditions for E to fall into the various standard types.

Under what conditions are all states of the process ergodic, and what is then the limiting distribution over states? (Manchester Dip.)

32. A game is played by three players, A_1, A_2, A_3, seated round a table: the winner being the first to get three heads in three independent tosses of an unbiased coin. A_1 begins: if he gets one tail in his three tosses he has another go. If he gets two tails the play passes to the player on his left. If he gets three tails the play passes to the player on his right. When the play passes to A_2 or A_3 they have the same rules. By considering a Markov chain of six states:

T_i: A_i is about to toss three coins ($i = 1, 2, 3$),

E_i: A_i has won ($i = 1, 2, 3$),

or otherwise, obtain the probability that the first player to toss wins. (Wales Dip.)

33. A Markov chain of four states has transition matrix

$$\begin{pmatrix} \frac{1}{2} & \frac{1}{2} & 0 & 0 \\ \frac{1}{4} & \frac{1}{2} & \frac{1}{4} & 0 \\ 0 & \frac{1}{4} & \frac{1}{2} & \frac{1}{4} \\ 0 & 0 & \frac{1}{2} & \frac{1}{2} \end{pmatrix}.$$

Find the stationary distribution. If the chain is in the first state (corresponding to the first row) find the expected number of transitions before it is next in the fourth state.

34. Use the method for finite Markov chains described in §4.5, based on characteristic vectors, to find the exact distribution at the sth trial for the Markov chain of the previous question, starting from a general initial distribution. In particular find the probability of being in the third state at the fifth trial, given that the chain was in the first state at the first trial (that is, after four transitions).

35. A chess club of N members $M_1, M_2, \ldots, M_N$ play a tournament in the following manner. M_1 plays M_2; the winner plays M_3; the winner of that game plays M_4 and so on. The winner of the first game against M_N plays M_1, and so on. The winner of the tournament is the person who first defeats his fellow members in $(N-1)$ successive games. If all players are equally good and draws are ignored find the expected number of plays in a tournament.

36. Set up a Markov chain formulation of the previous problem. If $N = 4$ and M_4 has probability p ($> \frac{1}{2}$) of beating any of the other players who are equally good, find the probability of M_4 winning the tournament.

37. The probability density of the lifetime of a piece of industrial equipment is $f(x)$. The equipment is inspected at unit intervals and classified in one of the states E_i. E_i ($i > 0$) means it has been inspected i times and is still working. E_0 means it has failed in the last interval, when it is replaced by a new piece. Show that the process on $\{E_i\}$ is Markov and find the transition probabilities in terms of $f(x)$.

38. A number, N, of points are placed around the perimeter of a circle. A particle at any one of these points has probability p of moving to the point to its right and probability $1-p = q$ of moving to the point to its left, irrespective of its previous movements. Show there exists a limiting stationary distribution, and find it.

A particle starts at a point A. It is said to have completed a 'tour' of the circle if the next time it reaches A it does so from the point to the other side of A from which it left A. Thus, if initially it moved to the left, it will need to return to A through the point to the right in order to complete a tour. Find the probability that the first return to A will constitute a tour and find the average duration of a tour.

39. A population of organisms consists initially of N_1 male and N_2 female organisms. In time δt any particular male is equally likely to mate with any particular female with probability $\lambda\delta t + o(\delta t)$, where λ is independent of time. Each mating immediately produces one offspring which is equally likely to be male or female. Obtain a differential equation for the probability generating function of the numbers n_1, n_2 of males and females present after time t.

Hence, or otherwise, show that for all time,

$$\mathscr{E}(n_1-n_2) = N_1-N_2, \quad \mathscr{D}^2(n_1-n_2) = \{\mathscr{E}(n_1+n_2)-(N_1+N_2)\}.$$

(Camb. Dip.)

40. A Markov chain with three states has transition matrix

$$P = \begin{bmatrix} q & p & 0 \\ 0 & q & p \\ p & 0 & q \end{bmatrix},$$

where $p+q = 1$. Show that if $0 < p < 1$ the chain is ergodic, with a limiting stationary distribution in which all states are equally probable.

Prove that

$$P^n = \begin{bmatrix} a_{1,n} & a_{2,n} & a_{3,n} \\ a_{3,n} & a_{1,n} & a_{2,n} \\ a_{2,n} & a_{3,n} & a_{1,n} \end{bmatrix},$$

where
$$a_{1,n}+\omega a_{2,n}+\omega^2 a_{3,n} = (q+p\omega)^n,$$

ω being a primitive cube root of unity.

If the states are assigned values 1, 2, 3, show that the autocorrelation function of the limiting stationary process is

$$\rho_n = a_{1,n}-\tfrac{1}{2}a_{2,n}-\tfrac{1}{2}a_{3,n}.$$

(Camb. Dip.)

41. Events occurring in time are observed not to be random but to be such that the occurrence of one event inhibits the occurrence of another event for a short period. It is therefore proposed to consider a model in which the probability that an event occurs in a short interval of time of length δt, given that the last event occurred at a time t before the commencement of the short interval, is

$$\lambda(1-e^{-\beta t})\,\delta t+o(\delta t).$$

This probability is independent of all previous occurrences other than the last. By finding the probability that no event occurs in an interval of length t after the occurrence of an event, or otherwise, show that if β is large the events occur at a rate $\lambda-\lambda^2/\beta$ approximately. (Camb. Dip.)

42. The sequence of numbers

$$x_1, x_2, x_3, \ldots, x_n, \ldots$$

represents the values of the density in a sequence of batches of material of equal size produced in the indicated order. Each x_n is independent of the others, and

$$p\{x_n = 9\} = p\{x_n = 11\} = p,$$

while

$$p\{x_n = 10\} = 1-2p.$$

A batch is either disposed of immediately, or put into store. It is disposed of immediately if its density is 10, or if it can be mixed with a batch already in store to give a mean density of 10; otherwise, it is put into store.

Find an approximation, valid for large N, for the expected number of batches in store, and the variance of this number, after N batches have been produced. (Camb. Dip.)

43. A simple model for population growth is to assume that each individual has a probability $\lambda\,\delta t+o(\delta t)$ of producing another in an interval δt of time independently of its previous history and other members of the population. Show that the logarithm of the mean number in the population increases linearly with time.

Another model supposes that the individuals exist in two forms A and B. Those of type A cannot produce offspring but have a probability $\nu\,\delta t+o(\delta t)$ of turning into type B, those of type B act as already described, producing always offspring of type A. Show that the logarithm of the mean of the total number of individuals still increases linearly with time, for long periods of growth, but that the rate is decreased by an amount $\lambda^2\tau+o(\tau)$, where τ is the mean lifetime spent as type A. (Camb. Dip.)

44. A slot machine works on inserting a penny. If the player wins, the penny is returned with an additional penny, otherwise the original penny is lost. The probability of winning is arranged to be $\frac{1}{2}$ (independently of previous plays) unless the previous play has resulted in a win, in which case the probability is $p < \frac{1}{2}$. Show that if the cost of maintaining the machine to the owner averages c pence per play, then the owner must arrange that $p < (1-3c)/2(1-c)$, $(c < \frac{1}{3})$, in order to make a profit in the long run. (Camb. Dip.)

45. In a radiation experiment, nuclear particles arrive randomly at a Geiger counter at rate λ per unit time. Each particle produces in the recording circuit of the counter a pulse of current which is of magnitude

$g(u)$ $(0 < u < \infty)$ at a time u after the arrival of the particle, so that the total current

$$x(t) = \int_0^t g(u)\,dN(t-u),$$

where $N(t)$ denotes the number of particles arriving during the interval $(0, t)$. By considering the cumulant generating functions of random variables of the form $g(u)\,[N(t-u+\delta)-N(t-u)]$, where δ is a small positive quantity, or otherwise, show that the cumulant generating function of $x(t)$, $\ln \mathscr{E}(\exp\theta x(t))$, tends to

$$\lambda \int_0^\infty \{e^{g(u)\theta}-1\}\,du \quad \text{when} \quad t \to \infty.$$

[It may be assumed that $\int_0^\infty |g(u)|\,du$ is finite and that

$$g(u) \to 0 \quad \text{when} \quad u \to \infty.]$$

Show also that for large t, the process $\{x(t)\}$ has autocorrelation function

$$\rho(\tau) = \int_0^\infty g(u)g(u+\tau)\,du \Big/ \int_0^\infty g^2(u)\,du. \qquad \text{(Camb. Dip.)}$$

46. In a simple birth-and-death process there is for any one individual a constant probability $\lambda\delta t + o(\delta t)$ of giving birth to a new individual in the time interval $(t, t+\delta t)$ and (independently of this) a constant probability $\mu\delta t + o(\delta t)$ of dying in $(t, t+\delta t)$, the probabilities for different individuals being mutually independent. Prove that the probability generating function of N_t, the population size at time t, given that $N_0 = n_0$, is

$$G(z, t) = \sum_{r=0}^{\infty} p(N_t = r)z^r = \left\{\frac{\mu(1-z)-(\mu-z)e^{-(\lambda-\mu)t}}{\lambda(1-z)-(\mu-z)e^{-(\lambda-\mu)t}}\right\}^{n_0}.$$

Hence, or otherwise, prove that $\mathscr{E}(N_t) = n_0 e^{(\lambda-\mu)t}$ and that the probability of the population ultimately becoming extinct is 1 $(\lambda \leqslant \mu)$ and $(\mu/\lambda)^{n_0}$ $(\lambda > \mu)$.

Prove that for $\lambda > \mu$,

$$p(N_t = r \mid D) = p(N_t = r)\,(\mu/\lambda)^{r-n_0},$$

where D denotes ultimate extinction of the population, and hence show that then $\mathscr{E}(N_t \mid D) = n_0 e^{(\mu-\lambda)t}$. (Camb. Dip.)

47. A process can be in either of two states A, B. Initially it is in state A, and changes of state take place at times

$$X_1, \quad X_1+Y_1, \quad X_1+Y_1+X_2, \quad X_1+Y_1+X_2+Y_2, \quad \ldots.$$

The random variables $X_1, Y_1, X_2, Y_2, \ldots$, representing successive lengths of time spent in a given state, are mutually independent and

$$p(X_i < x) = G(x), \quad p(Y_i \leqslant y) = H(y) \quad (i = 1, 2, \ldots).$$

Let

$$G_n(x) = p\left(\sum_{i=1}^{n} X_i < x\right), \quad H_n(y) = p\left(\sum_{i=1}^{n} Y_i \leqslant y\right) \quad (n = 1, 2, \ldots);$$

$$G_0(x) \equiv 1, \quad \text{and} \quad H_0(y) = 1 \;\text{ if }\; y \geqslant 0, \quad H_0(y) = 0 \;\text{ if }\; y < 0.$$

Let $\beta(t)$ be the total time spent in the state B during the interval $(0, t)$. Show that

$$p\{\beta(t) \leqslant x\} = \sum_{n=0}^{\infty} H_n(x)\,[G_n(t-x) - G_{n+1}(t-x)].$$

For $G(x) = 1 - e^{-\lambda x}$, $H(y) = 1 - e^{-\mu y}$, obtain

$$p\{\beta(t) \leqslant x\} = e^{-\lambda(t-x)}\left[1 + \{\lambda\mu(t-x)\}^{\frac{1}{2}} \int_0^x e^{-\mu y} y^{-\frac{1}{2}} I_1[2\{\lambda\mu(t-x)y\}^{\frac{1}{2}}]\,dy\right],$$

where
$$I_1(z) = \sum_{r=0} \frac{(\frac{1}{2}z)^{2r+1}}{r!\,(r+1)!}.$$ (Camb. Dip.)

48. $\{X_t\}$ $(t = 0, 1, 2, \ldots)$ is a time-homogeneous Markov chain whose states are the positive integers. The probabilities that $X_{t+n} = j$ and that $X_{t+n} = j$, $X_{t+n'} \neq j$ if $1 \leqslant n' < n$, each conditional on $X_t = i$, are respectively $p_{ij}^{(n)}$ and $f_{ij}^{(n)}$ $(n = 1, 2, 3, \ldots)$. Show that

$$F_{ii}(z) = 1 - \left\{1 + \sum_{n=1}^{\infty} z^n p_{ii}^{(n)}\right\}^{-1} \quad (|z| < 1),$$

where
$$F_{ii}(z) = \sum_{n=1}^{\infty} z^n f_{ii}^{(n)},$$

and hence that the state i is persistent if and only if the series $\sum_n p_{ii}^{(n)}$ diverges. Explain (without proof) how the mean recurrence time of a persistent state is related to the behaviour of $p_{ii}^{(n)}$ as $n \to \infty$.

For a chain with two states 1 and 2,

$$p_{11}^{(1)} = 1-a, \quad p_{22}^{(1)} = 1-b \quad (a, b > 0).$$

By obtaining $F_{11}(z)$, or otherwise, determine the mean and variance of the distribution of the recurrence time of the state 1. (Camb. Dip.)

49. A traffic light has a constant probability $\lambda\,dt$ of changing to green after being red or to red after being green in any infinitesimal time interval dt, so that a car arriving at a random instant has a probability of $\frac{1}{2}$ of passing through without waiting and a probability element $\frac{1}{2}\lambda \exp(-\lambda w)\,dw$ of waiting a time w, where $w > 0$. Two cars approach the traffic light the second one arriving at a time interval a after the first. Find the joint distribution of their waiting times, w_1 and w_2, and hence the distribution of $w_2 - w_1$.

If the two cars travel always with the same speed between traffic lights and are following the same route, can you say anything about the distribution of the time by which the second car is behind the first, after they have passed a large number of traffic lights operating independently with the same coefficient λ? (Camb. Dip.)

50. The n states of a time-homogeneous Markov chain are represented by the integers $0, 1, 2, \ldots, n-1$. p_{ij} is the probability that the chain is in state j at trial $k+1$, given that it was in state i at trial k. Derive a set of

n linear equations satisfied by x_j, the probability, starting from state j at the first trial, of reaching, after the first trial, the state 0 *before* the state $n-1$ ($j = 0, 1, ..., n-1$).

A time-homogeneous Markov chain with two states has $p_{00} = p_0$, $p_{11} = p_1$. From a realization of this chain a new chain is constructed by considering non-overlapping pairs of integers, calling the pair 01, state 0; and the pair 10, state 1; the pairs 00, 11 being ignored. Prove that the new chain has $p_{00} = p_{11} = (1+p_0+p_1)^{-1}$.

Show how, from any realization of this new chain, an *independent* sequence of the integers 0 and 1 may be derived. (Camb. Dip.)

BIBLIOGRAPHY

ANDERSON, T. W. (1958). *An Introduction to Multivariate Statistical Analysis.* New York: John Wiley and Sons Inc.

BARTLETT, M. S. (1956). *An Introduction to Stochastic Processes.* Cambridge University Press.

BAYES, T. (1958). Essay towards solving a problem in the doctrine of chances. *Biometrika,* **45**, 293–315. (Reproduction of 1763 paper.)

BRAITHWAITE, R. B. (1953). *Scientific Explanation.* Cambridge University Press.

BROCKMEYER, E., HALSTRØM, H. L. and JENSEN, A. (1948). *The Life and Works of A.K. Erlang.* Copenhagen Telephone Company.

CARNAP, R. (1950). *Logical Foundations of Probability.* London: Routledge and Kegan Paul Ltd.

CHUNG, KAI LAI (1960). *Markov Chains with Stationary Transition Probabilities.* Berlin: Springer-Verlag.

COX, D. R. (1962). *Renewal Theory.* London: Methuen and Co. Ltd.

COX, D. R. and SMITH, W. L. (1961). *Queues.* London: Methuen and Co. Ltd.

CRAMÉR, H. (1946). *Mathematical Methods of Statistics.* Princeton University Press.

DAVID, F. N. and BARTON, D. E. (1962). *Combinatorial Chance.* London: Charles Griffin and Co. Ltd.

DOOB, J. L. (1953). *Stochastic Processes.* New York: John Wiley and Sons Inc.

DYNKIN, E. B. (1960). *Theory of Markov Processes.* Oxford: Pergamon Press. (Translation from the Russian.)

FADDEEVA, V. N. (1959). *Computational Methods of Linear Algebra.* New York: Dover Publications Inc. (Translation from the Russian.)

FELLER, W. (1957). *An Introduction to Probability Theory and Its Applications,* Vol. 1. New York: John Wiley and Sons Inc.

FINETTI, B. DE (1937). La prévision: ses lois logiques, ses sources subjectives. *Annales de l'Institut Henri Poincaré,* 7, 1–68.

FRAZER, R. A., DUNCAN, W. J. and COLLAR, A. R. (1938). *Elementary Matrices.* Cambridge University Press.

GNEDENKO, B. V. (1962). *The Theory of Probability.* New York: Chelsea Publishing Co.

GNEDENKO, B. V. and KOLMOGOROV, A. N. (1954). *Limit Distributions for Sums of Independent Random Variables.* Reading: Addison-Wesley Publishing Co. Inc. (Translation from the Russian.)

GOOD, I. J. (1950). *Probability and the Weighing of Evidence.* London: Charles Griffin and Co. Ltd.

HALMOS, P. R. (1950). *Measure Theory.* New York: D. van Nostrand Co. Inc.

HARVARD COMPUTATION LABORATORY STAFF (1955). *Tables of the Cumulative Binomial Probability Distribution.* Cambridge: Harvard University Press.

JEFFREYS, H. (1961). *Theory of Probability*, 3rd edition. Oxford: Clarendon Press.

KAC, M. (1959). *Statistical Independence in Probability, Analysis and Number Theory.* New York: John Wiley and Sons Inc.

KEMENY, J. G. and SNELL, J. L. (1960). *Finite Markov Chains.* Princeton: D. van Nostrand Co. Inc.

KENDALL, D. G. (1951). Some problems in the theory of queues. *J. roy. statist. Soc.* B, **13**, 151–185.

KENDALL, M. G. and MORAN, P. A. P. (1963). *Geometrical Probability.* London: Charles Griffin and Co. Ltd.

KENDALL, M. G. and BABINGTON SMITH, B. (1954). *Tables of Random Sampling Numbers.* Cambridge University Press.

KERRICH, J. E. (1946). *An Experimental Introduction to the Theory of Probability.* Copenhagen: Einar Munksgaard.

KEYNES, J. M. (1921). *A Treatise on Probability.* London: MacMillan and Co. Ltd.

KHINTCHINE, A. Y. (1960). *Mathematical Methods in the Theory of Queueing.* London: Charles Griffin and Co. Ltd. (Translation from the Russian.)

KOLMOGOROV, A. N. (1956). *Foundations of the Theory of Probability.* New York: Chelsea Publishing Co. (Translation of original 1933 German edition.)

LAPLACE, P. S. (1951). *A Philosophical Essay on Probabilities.* London: Constable and Co. Ltd. (Translation of 1819 French edition.)

LINDLEY, D. V. and MILLER, J. C. P. (1961). *Cambridge Elementary Statistical Tables.* Cambridge University Press.

LOÈVE, M. (1960). *Probability Theory*, 2nd edition. Princeton: D. van Nostrand Co. Inc.

MACMAHON, P. A. (1916). *Combinatory Analysis.* Cambridge University Press.

MUNROE, M. E. (1953). *Introduction to Measure and Integration.* Cambridge: Addison-Wesley Publishing Co. Inc.

NATIONAL BUREAU OF STANDARDS (U.S.A.) (1953). *Tables of Normal Probability Functions.* Washington D.C.

NEUMANN, J. VON and MORGENSTERN, O. (1947). *Theory of Games and Economic Behaviour.* Princeton University Press.

PARZEN, E. (1962). *Stochastic Processes.* San Francisco: Holden-Day Inc.

PITT, H. R. (1963). *Integration, Measure and Probability.* Edinburgh and London: Oliver and Boyd.

RAIFFA, H. and SCHLAIFER, R. (1961). *Applied Statistical Decision Theory.* Boston: Harvard University Graduate School of Business Administration.

RAMSEY, F. P. (1931). *The Foundations of Mathematics*. London: Routledge and Kegan Paul Ltd.

RAND CORPORATION (1955). *A Million Random Digits with* 100,000 *Normal Deviates*. Glencoe (Illinois): Free Press.

REICHENBACH, HANS (1949). *The Theory of Probability*. Berkeley and Los Angeles: University of California Press.

RENYI, A. (1962). *Wahrscheinlichkeitsrechnung*. Berlin: Veb Deutscher Verlag der Wissenschaften.

RIORDAN, J. (1958). *An Introduction to Combinatorial Analysis*. New York: John Wiley and Sons Inc.

ROMIG, H. G. (1953). 50–100 *Binomial Tables*. New York: John Wiley and Sons Inc.

SAVAGE, L. J. (1954). *The Foundations of Statistics*. New York: John Wiley and Sons Inc.

WHITWORTH, W. A. (1901). *Choice and Chance*. Cambridge: Deighton Bell and Co.

WOLD, HERMAN (1954). *Random Normal Deviates*. Cambridge University Press.

SUBJECT INDEX

INDEX OF NOTATIONS